Student Solutions

Calculus Concepts, An Informal Approach to the Mathematics of Change

FIFTH EDITION

Donald R. LaTorre
Clemson University

John W. Kenelly
Clemson University

Sherry Biggers
Clemson University

Laurel R. Carpenter
Charlotte, Michigan

Iris B. Reed
Clemson University

Cynthia R. Harris
Reno, Nevada

Prepared by

Laurel Carpenter
Charlotte, Michigan

Sherry Biggers
Clemson University

Australia • Brazil • Japan • Korea • Mexico • Singapore • Spain • United Kingdom • United States

© 2012 Brooks/Cole, Cengage Learning

ALL RIGHTS RESERVED. No part of this work covered by the copyright herein may be reproduced, transmitted, stored, or used in any form or by any means graphic, electronic, or mechanical, including but not limited to photocopying, recording, scanning, digitizing, taping, Web distribution, information networks, or information storage and retrieval systems, except as permitted under Section 107 or 108 of the 1976 United States Copyright Act, without the prior written permission of the publisher.

For product information and technology assistance, contact us at
**Cengage Learning Customer & Sales Support,
1-800-354-9706**

For permission to use material from this text or product, submit all requests online at **www.cengage.com/permissions**
Further permissions questions can be emailed to **permissionrequest@cengage.com**

ISBN-13: 978-0-538-73541-4
ISBN-10: 0-538-73541-4

Brooks/Cole
20 Channel Center Street
Boston, MA 02210
USA

Cengage Learning is a leading provider of customized learning solutions with office locations around the globe, including Singapore, the United Kingdom, Australia, Mexico, Brazil, and Japan. Locate your local office at: **www.cengage.com/global**

Cengage Learning products are represented in Canada by Nelson Education, Ltd.

To learn more about Brooks/Cole, visit **www.cengage.com/brookscole**

Purchase any of our products at your local college store or at our preferred online store **www.cengagebrain.com**

Printed in the United States of America
1 2 3 4 5 6 7 15 14 13 12 11

Contents

1. Ingredients of Change: Functions and Limits 1
2. Describing Change: Rates 42
3. Determining Change: Derivatives 76
4. Analyzing Change: Applications of Derivatives 109
5. Accumulating Change: Limits of Sums and the Definite Integral..160
6. Analyzing Accumulated Change: Integrals in Action................207
7. Ingredients of Multivariable Change: Models, Graphs, Rates......250
8. Analyzing Multivariable Change..282

Chapter 1   Ingredients of Change: Functions and Limits

Section 1.1   Functions—Four Representations (pages 8–12)

1. a. A scatter-plot is a ⬚graphical⬚ representation.
    b. Input is shown along the horizontal axis. Input is ⬚the number of years since 9/1/2000⬚, and the units of measure are ⬚years⬚.
        Output is shown along the vertical axis. Output is ⬚the performance of an investment⬚ measured in ⬚dollars⬚.
    c. The relation is ⬚a function⬚ because only one performance amount is possible at any given time.
    d. t years

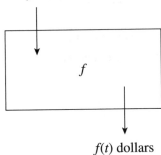

$f(t)$ dollars

3. a. The statement is a ⬚verbal⬚ representation.
    b. Input is ⬚the fraction of certain relatives who have had thyroid cancer⬚. The fraction used as input is ⬚unit-less⬚.
        Output is ⬚the chance of developing thyroid cancer⬚ measured in ⬚percentage points⬚.
    c. The relation is ⬚a function⬚ because a specific percentage risk is associated with each fraction.
    d. f

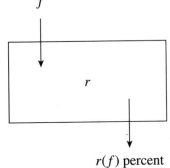

$r(f)$ percent

5. a. A formula is an ⬚algebraic⬚ representation.
    b. The input variable is n. Input is ⬚the number of correctly spelled words⬚, and units of measure are ⬚words⬚.
        The output of the formula is $g(n)$. Output is a ⬚raw score⬚ which is ⬚unit-less⬚.

c. The relation is a function because a certain number of words spelled correctly results in only one raw score.

d.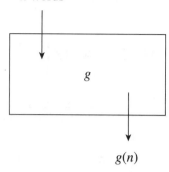

7. a. A table of a data is a numerical representation.
   b. The input is shown in the left column. Input is the diameter of the raw loaf, and the units of measure are inches.
   The output is shown in the right column. Output is baking time measured in minutes.
   c, d. The relation is not a function because baking time for a given diameter can vary by 10- to 15- minutes.

9. a. A statement is a verbal representation.
   b. Input is the calendar year, and the units of measure are years.
   Output is the pre-season poll ranking and is unit-less.
   c. The relation is a function because there is only one ranking for each team in the Sports Illustrated pre-season poll.
   d.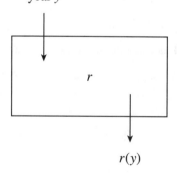

11. $s(5) = 22$
    $s(10) = 38$

13. $t(-11) = -638$
    $t(4) = -68$

15. $r(4) = 10.4976 \rightarrow \approx \boxed{10.498}$

    $r(-0.5) \approx \boxed{0.745}$

17. $t(3) \approx \boxed{22.225}$

    $t(0.2) \approx \boxed{16.378}$

19. Solving $t(x) = 10$ yields $x = \boxed{-1}$ and $x = \boxed{1.6}$

    Solving $t(x) = 15$ yields $x \approx \boxed{-1.340}$ and $x \approx \boxed{1.940}$

21. Solving $s(t) = 18$ yields $t \approx \boxed{-4.786}$ and $t \approx \boxed{2.786}$

    Solving $s(t) = 0$ yields $t \approx \boxed{-4.641}$ and $t \approx \boxed{-0.359}$

23. Solving $r(x) = 9.4$ yields $x \approx \boxed{3.537}$

    Solving $r(x) = 30$ yields $x \approx \boxed{5.511}$

25. Solving $t(n) = 7.5$ yields $n \approx \boxed{1.386}$

    Solving $t(n) = 1.8$ yields $n \approx \boxed{-2.599}$

27. $f(x) = 7$ is an $\boxed{\text{output}}$ value

    Solving $f(x) = 7$ yields $x \approx \boxed{4.953}$

29. $t = 15$ is an $\boxed{\text{input}}$ value

    $A(15) \approx \boxed{57,857.357}$

31. $g(x) = 247$ is an $\boxed{\text{output}}$ value

    Solving $g(x) = 247$ yields $x = \boxed{-13}$ and $x = \boxed{5}$

33. $x = 10$ is an $\boxed{\text{input}}$ value

    $m(10) \approx \boxed{2.429}$

35. a. An input value of 3 represents June and an output value of 12.90 represents 12.90 hundred gallons of motor oil.

    $\boxed{\text{In June, the local stock car racing team used 1290 gallons of motor oil.}}$

b. October is represented by an input value of $t = 10 - 3 = 7$ and 1,520 gallons is represented by the corresponding output value of $\frac{1520}{100} = 15.20$.

$\boxed{g(7) = 15.20}$

37. a. An input value of 0 represents the year 2003 and an output value of 61.5 represents 61.5 million pet dogs.

$\boxed{\text{In 2003, there were 61.5 million pet dogs in the United States.}}$

b. The year 2008 is represented by an input value of $t = 2008 - 2003 = 5$ and 66.3 million is represented by an output value of 66.3.

$\boxed{p(5) = 66.3}$

39. a. The year 2000 is represented by an input value of $t = 2000 - 2001 = -1$.
The output is calculated as $d(-1) \approx 1.023$ → $\boxed{\$1.02}$ (Round to the nearest cents.)
The year 2010 is represented by an input value of $t = 2010 - 2001 = 9$.
The output is calculated as $d(9) \approx 0.793$ → $\boxed{\$0.79}$ (Round to the nearest cents.)

b. Solving $d(t) = 0.80$ yields $t \approx 8.696$
Converting 8.696 into years and months gives 8 years and $0.696 \cdot 12 \approx 8.352$ months since the end of 2001.
$\boxed{\text{September, 2010}}$
Solving $d(t) = 0.75$ yields $t \approx 10.870$
Converting 10.870 into years and months gives 10 years and $0.870 \cdot 12 \approx 10.440$ months since the end of 2001.
$\boxed{\text{November, 2012}}$

41. a. A depth of 75 feet is represented by an input value of $x = 75$.
The output is calculated as $t(75) \approx 39.175$ → $\boxed{39 \text{ minutes}}$ (Rounded to the nearest minute.)
A depth of 95 feet is represented by an input value of $x = 95$.
The output is calculated as $t(95) \approx 23.036$ → $\boxed{23 \text{ minutes}}$ (Rounded to the nearest minute.)

b. A maximum dive time of 20 minutes is represented as an output value of 20.
Solving $t(x) = 20$ yields $x \approx 100.323$ → $\boxed{100 \text{ feet}}$ (Rounded to the nearest feet.)

43. y is not a function because a vertical line can be drawn that intersects the graph at more than one point.

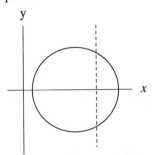

45. y is a function because any vertical line drawn through the graph intersects the graph at no more than one point.

47. y is a function because any vertical line drawn through the graph intersects the graph at no more than one point.

Section 1.2 Function Behavior and End Behavior Limits (pages 19–22)

1. increasing, concave down

3. decreasing for $x < c$
   increasing for $x > c$
   concave up for all x shown

5. increasing for $x < b$
   decreasing for $x > b$
   concave down for $x < 0$ and $a < x$
   concave up for $0 < x < a$

7. a. increasing for all t shown
   b. The concavity of N appears to change from concave up to concave down when $t \approx 4.7$.

9. A graph of $f(x) = 45.183(0.831^x)$ reveals the following behavior:

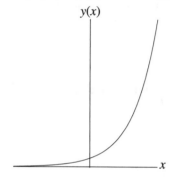

a. decreasing
b. concave up

11. A graph of $y(x) = 1.5^x$ suggests the following behavior:

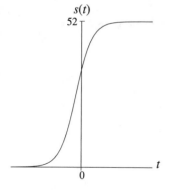

a. As x decreases without bound, the output values of y approach zero.
   As x increases without bound, the output values of y increase without bound.
b. $\lim_{x \to -\infty} y(x) = 0$

   $\lim_{x \to \infty} y(x) = \infty$

c. The horizontal axis $\boxed{y = 0}$ is a horizontal asymptote.

13. A graph of $s(t) = \dfrac{52}{1 + 0.5e^{-0.9t}}$ suggests the following behavior:

a. As t decreases without bound, the output values of s approach zero.
   As t increases without bound, the output values of s approach a limiting value of 52.

b. $\lim_{t \to -\infty} s(t) = 0$

$\lim_{t \to \infty} s(t) = 52$

c. The horizontal axis $\boxed{y=0}$ and a horizontal line at $\boxed{y=52}$ are the two horizontal asymptotes.

15. A graph of $n(x) = 4x^2 - 2x + 12$ suggests the following behavior:

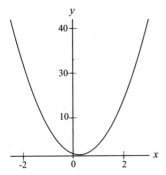

a. The output values of n increase without bound as x increases (or decreases) without bound.
b. $\lim_{x \to \pm\infty} n(x) = \infty$
c. There are $\boxed{\text{no horizontal asymptotes}}$ because n is unbounded.

17. A graph of $f(x) = 5xe^{-x}$ suggests the following behavior:

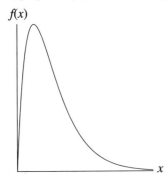

a. As x increases without bound, the output values of f approach zero from above.
b. $\lim_{x \to \infty} f(x) = 0$
c. The horizontal axis $\boxed{y=0}$ is the horizontal asymptote.

19.

$x \to \infty$	$1 - 0.5^x$
5	0.9688
10	0.9990
20	0.9999990
40	0.9999999...
$\lim_{x \to \infty}(1 - 0.5^x) \approx 1.00$	

21.

$x \to \infty$	$(1+x^{-1})^x$
1,000	2.71692
10,000	2.71815
100,000	2.71827
1,000,000	2.71828
10,000,000	2.71828
$\lim_{x \to \infty} \left[(1+x^{-1})^x\right] \approx 2.718$	

(Round-off error occurs by $x = 10^{13}$. By $x = 10^{14}$ a TI84 reports $(1+x^{-1})^x = 1$.)

23.

$x \to \infty$	-1.5^x
5	-7.59
10	-57.67
20	-3,325.26
40	-11,057,332.32
$\lim_{x \to \infty}(-1.5^x) = -\infty$	

(Overflow occurs by $x = 600$.)

As x increases without bound, -1.5^x decreases without bound. The limit $\boxed{\text{does not exist}}$.

25. a.

$t \to \infty$	$N(t)$
5	1761.02
10	3003.21
15	3015.92
20	3016.00
25	3016.00
$\lim_{t \to \infty}[N(t)] \approx 3,016$	

b. The horizontal line at $\boxed{y = 3016}$ is a horizontal asymptote for N.

c. The model predicts that a total of 3,016 deaths occurred among members of the U.S. Navy as a result of influenza during the epidemic of 1918.

27. a.

$x \to \infty$	$f(x)$
5	77.797
10	67.010
20	61.088
40	60.026
80	60.000
160	60.000
$\lim_{x \to \infty}[f(x)] \approx 60.0$	

b. The horizontal line at $y = 60.0$ is a horizontal axis of f.

c. The model predicts that eventually the number of farms with milk cows will decrease to and stay near 60 thousand farms.

Section 1.3 Limits and Continuity (pages 30–31)

1. a. $\lim_{x \to 2.2^-} f(x) = \boxed{3}$

   b. $\lim_{x \to 2.2^+} f(x) = \boxed{3}$

   c. $\lim_{x \to 2.2} f(x) = \boxed{3}$

   d. $f(2.2) = \boxed{3}$

   e. yes; The function f is continuous at $x = 2.2$, because the limit exists and is equal to the function value at $x = 2.2$.

3. a. $\lim_{t \to 1^-} m(t) = \boxed{1.5}$

   b. $\lim_{t \to 1^+} m(t) = \boxed{1.5}$

   c. $\lim_{t \to 1} m(t) = \boxed{1.5}$

   d. $m(1) = \boxed{2}$

   e. no; The function m is not continuous at $t = 1$ because even though the limit exists at $t = 1$ it does not equal the output value of the function for $t = 1$.

5. a. $\lim_{x \to 6^-} f(x) = \boxed{-\infty}$

   b. $\lim_{x \to 6^+} f(x) = \boxed{-1}$

   c. $f(6) = \boxed{-1}$

   d. no; The function f is not continuous at $x = 6$ because the left portion of f approaches a vertical asymptote as x approaches 6 from the left and the right portion of f approaches -1 as x approaches 6 from the right.

7. a. $\lim_{x \to 0^-} g(x) = \boxed{-3}$

   b. $\lim_{x \to 0^+} g(x) = \boxed{-3}$

   c. $g(0) = \boxed{-3}$

   d. yes; The function g is continuous at $x = 0$ because the limit exists ($\lim_{x \to 0^-} g(x) = \lim_{x \to 0^+} g(x)$) and is equal to the output value at $x = 0$.

9. a. $\lim_{x \to 8} g(x) = \boxed{12}$

   b. $\lim_{x \to \infty} g(x) = \boxed{\infty}$

11.

$x \to 3^-$	$\dfrac{1}{x-3}$
2.9	-10
2.99	-100
2.999	-1000
2.9999	-10,000
$\lim\limits_{x \to 3^-}\left(\dfrac{1}{x-3}\right) = -\infty$	

$x \to 3^+$	$\dfrac{1}{x-3}$
3.1	10
3.01	100
3.001	1000
3.0001	10,000
$\lim\limits_{x \to 3^+}\left(\dfrac{1}{x-3}\right) = \infty$	

$\lim\limits_{x \to 3}\left(\dfrac{1}{x-3}\right)$ $\boxed{\text{does not exist}}$ because the limit from the right does not exist the limit from the left.

13.

$x \to 5^-$	$\dfrac{2x-10}{x-5}$
4.9	2
4.99	2
4.999	2
4.9999	2
$\lim\limits_{x \to 5^-}\left(\dfrac{2x-10}{x-5}\right) = 2$	

$x \to 5^+$	$\dfrac{2x-10}{x-5}$
3.1	2
3.01	2
3.001	2
3.0001	2
$\lim\limits_{x \to 5^+}\left(\dfrac{2x-10}{x-5}\right) = 2$	

$\lim\limits_{x \to 5}\left(\dfrac{2x-10}{x-5}\right) = \boxed{2}$ because the limit from the right is equal to the limit from the left.

15.

$h \to 0^-$	$\dfrac{(3+h)^2 - 3^2}{h}$
-0.1	5.9
-0.01	5.99
-0.001	5.999
-0.0001	5.9999
$\lim\limits_{h \to 0^-}\left(\dfrac{(3+h)^2 - 3^2}{h}\right) = 6$	

$h \to 0^+$	$\dfrac{(3+h)^2 - 3^2}{h}$
0.1	6.1
0.01	6.01
0.001	6.001
0.0001	6.0001
$\lim\limits_{h \to 0^+}\left(\dfrac{(3+h)^2 - 3^2}{h}\right) = 6$	

$\lim\limits_{h \to 0}\left(\dfrac{(3+h)^2 - 3^2}{h}\right) = \boxed{6}$ because the limit from the right is equal to the limit from the left.

17. $\lim\limits_{x \to 0} 9 = \boxed{9}$ by the Constant Rule

19. $\lim\limits_{t \to 3}(6g(t)) = 6 \cdot \lim\limits_{t \to 3} g(t)$ by the Constant Multiplier Rule

$\qquad\qquad\qquad = 6 \cdot 5 \qquad\qquad \lim\limits_{t \to 3} g(t) = 5$ is given

$\qquad\qquad\qquad = \boxed{30}$

21. $\lim\limits_{x \to 0.1}[f(x) - g(x)] = \lim\limits_{x \to 0.1} f(x) - \lim\limits_{x \to 0.1} g(x)$ by the Difference Rule

$\qquad\qquad\qquad\qquad = 6 - 3 \qquad\qquad \lim\limits_{x \to 0.1} f(x) = 6$ and $\lim\limits_{x \to 0.1} g(x) = 3$ are given

$\qquad\qquad\qquad\qquad = \boxed{3}$

23. $\lim\limits_{t \to 3}(4t - 5) = 4 \cdot 3 - 5$ by the Replacement Rule

$\qquad\qquad\qquad = \boxed{7}$

25. $\lim\limits_{x \to -2}(x^2 - 4x + 4) = (-2)^2 - 4(-2) + 4$ by the Replacement Rule

$\qquad\qquad\qquad\qquad = \boxed{16}$

27. $\lim\limits_{m \to 0} \dfrac{m}{m^2 + 4m} = \lim\limits_{m \to 0} \dfrac{m}{m(m + 4)}$ by factoring the denominator

$\qquad\qquad\qquad = \lim\limits_{m \to 0} \dfrac{1}{m + 4}$ by the Cancellation Rule

$\qquad\qquad\qquad = \dfrac{1}{0 + 4}$ by the Replacement Rule

$\qquad\qquad\qquad = \boxed{\dfrac{1}{4}}$ (or 0.25)

29. $\lim\limits_{t \to 4} \dfrac{t^2 - 4t}{t - 4} = \lim\limits_{t \to 4} \dfrac{t(t - 4)}{t - 4}$ by factoring the numerator

$\qquad\qquad\qquad = \lim\limits_{t \to 4} t$ by the Cancellation Rule

$\qquad\qquad\qquad = \boxed{4}$ by the Replacement Rule

31. $\lim\limits_{h \to 0} \dfrac{(5 + h)^2 - 5^2}{h} = \lim\limits_{h \to 0} \dfrac{h(10 + h)}{h}$ by expanding and then factoring the numerator

$\qquad\qquad\qquad = \lim\limits_{h \to 0}(10 + h)$ by the Cancellation Rule

$\qquad\qquad\qquad = 10 + 0$ by the Replacement Rule

$\qquad\qquad\qquad = \boxed{10}$

33. a. $\lim\limits_{x \to -1^-} f(x) = \lim\limits_{x \to -1^-} x^2$ using the $x < -1$ part of f

$\qquad\qquad\qquad = (-1)^2$ by the Replacement Rule

$\qquad\qquad\qquad = \boxed{1}$

b. $\lim_{x \to -1^+} f(x) = \lim_{x \to -1^+} 1$ using the $x \geq -1$ part of f

   $= \boxed{1}$ by the Constant Rule

c. $f(-1) = \boxed{1}$ using the $x \geq -1$ part of f

d. $\boxed{\text{yes}}$; The function f is continuous at $x = -1$, because the limit exists and is equal to the output value at $x = -1$.

35. a. $\lim_{x \to 2^-} f(x) = \lim_{x \to 2^-} 10x^{-1}$ using the $x < 2$ part of f

    $= 10(2)^{-1}$ by the Replacement Rule

    $= \boxed{5}$

b. $\lim_{x \to 2^+} f(x) = \lim_{x \to 2^+} (4x - 3)$ using the $x \geq 2$ part of f

   $= 4 \cdot 2 - 3$ by the Replacement Rule

   $= \boxed{5}$

c. $f(2) = \boxed{5}$ using the $x \geq 2$ part of f

d. $\boxed{\text{yes}}$; The function f is continuous at $x = 2$, because the limit exists and is equal to the output value at $x = 2$.

Section 1.4 Linear Functions and Models (pages 42–45)

1. a. $a = \boxed{0.3}$ (when a and b represent the constants in the function $f(x) = ax + b$)

   b. $\dfrac{\text{output units of measure}}{\text{input units of measure}} = \dfrac{\text{dollars}}{\text{year}}$

   $\boxed{\text{The cost to rent a newly released movie is increasing by \$0.30 per year.}}$

   c. $f(0) = b = \boxed{5}$

   $\boxed{\text{In 2010, the cost to rent a newly released movie was \$5.}}$

3. a. $a = \boxed{2}$ (when a and b represent the constants in the function $f(x) = ax + b$)

   b. $\dfrac{\text{output units of measure}}{\text{input units of measure}} = \dfrac{\text{thousand dollars}}{\text{hundred units}}$

   $\boxed{\text{The profit is increasing by 2 thousand dollars per hundred units.}}$

   c. $f(0) = b = \boxed{-4.5}$

   $\boxed{\text{When no units are sold the profit is -\$4.5 thousand.}}$

5. a. $a = \boxed{100}$ (when a and b represent the constants in the function $f(x) = ax + b$)

   b. $\dfrac{\text{output units of measure}}{\text{input units of measure}} = \dfrac{\text{feet}}{\text{dollar}}$

The production of wire is increasing by 100 feet per dollar spent for raw materials.

c. $f(0) = b = \boxed{0}$
When no money is spent for raw materials, no wire is produced.

7. rate of change = $a = 0.3$ dollars per toy, initial output = $b = 50$ dollars
$\boxed{C(x) = 0.3x + 50 \text{ dollars gives the cost to produce } x \text{ toys.}}$

9. rate of change = $a = 0.25$ inches per hour, initial output = $b = 0$ inches
$\boxed{S(h) = 0.25h \text{ inches gives the amount of snow that fell where } h \text{ is the number of hours since midnight, } 0 \leq h \leq 15.5.}$

11. rate of change = $a = 4.75$ square feet per minute, initial output = $b = -6$ square feet
$\boxed{F(t) = 4.75t - 6 \text{ square feet gives the amount of usable fabric sheeting manufactured in } t \text{ minutes.}}$

13. a. The graph of s shows that s is $\boxed{\text{increasing}}$ for all t shown.
   b. Using points (2002, 0.10) and (2006, 0.20) estimated from the graph of s yields

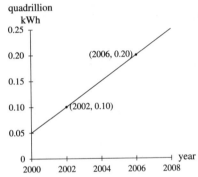

$$\frac{0.20 - 0.10}{2006 - 2002} \approx \boxed{0.025 \text{ quadrillion kWh per year}}$$

Retail sales of electricity were increasing by 0.025 quadrillion kWh per year between 2000 and 2008.

   c. Using the graph of s to estimate the output corresponding to 2005 yields

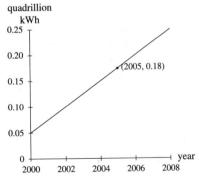

$s(2005) \approx \boxed{0.175 \text{ quadrillion kWh}}$

(The answer key in the book reports this estimate rounded to 0.18.)
In 2005, retail sales of electricity reached 0.18 quadrillion kWh.

15. a. Using points (1, 1760) and (3, 1830) estimated from the graph yields

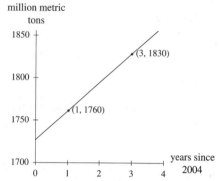

$$\frac{1830-1760}{3-1} \approx 35 \text{ million metric tons per year}$$

Carbon dioxide emission were increasing by an average of 35 million metric tons per year between 2004 and 2008.

b. Emissions are increasing. The positive slope reflects increasing emissions.

17. a. $a = $ 5.64 million users per year (where a and b are the constants of the linear model)
b. One possible figure:

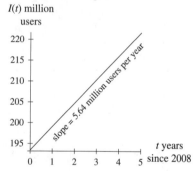

Other figures can be drawn using different scales for the vertical axis.

c. $I(0) = b = $ 193.44 million users

In 2008 there were 193.44 million Internet users in the United States.

d. (The answer key in the book correctly states that the result in part c was found using interpolation. However, the information stated in the activity is insufficient to form a conclusion because the data interval was not given.)

19. a. $\dfrac{41.9 - 59.9}{2006 - 1960} \approx $ −0.391 births per thousand women per year

b. $B(x) \approx -0.391x + 59.9$ births per thousand women per thousand is the birth rate for women aged 15 – 19 where x is the number of years since 1960, data from 1960 through 2006.

c. $B(52) \approx $ 39.552 births per thousand women per year

21. a. The scatter-plot suggests an increasing linear pattern.
    b. $D(t) \approx 1.023t + 78.360$ million barrels gives the daily world demand for oil where t is the number of years since 2000, data from $4 \leq t \leq 9$.
    c. $D(15) \approx \boxed{93.709 \text{ million barrels}}$ – assuming that demand continues to increase and can be met according to the established pattern

23. a. $P(t) = -28t + 45.2$ thousand pounds gives the inventory of peaches at a packing plant during a one hour period, data from $0 \leq t \leq 1$.
    b. $P(0.5) = \boxed{31.2 \text{ thousand pounds}}$
    c. $P(1) = \boxed{17.2 \text{ thousand pounds}}$

25. a. $m(x) \approx 3.683x + 125.076$ million users gives the number of people in North America who use email as a part of their job where x is the number of years since 2005, data from $0 \leq x \leq 10$.
    b. $a \approx \boxed{3.683 \text{ million users per year}}$ (where a and b are the constants of the linear model)
    c. $m(8) \approx \boxed{154.539 \text{ million users}}$
       2013 is outside the data input interval, so the estimate for $m(8)$ is $\boxed{\text{extrapolation}}$.

27. Answers might vary but should be similar to the following:
    The process of interpolation uses the model to calculate results within the input interval of the data while the process of extrapolation uses the model to calculate results outside the interval of the data.

Section 1.5 Exponential Functions and Models (pages 53–56)

1. (The figure in the activity is not correct, it should look like the following figure.)

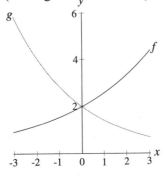

   a. The blue graph is $\boxed{\text{increasing}}$.
      The black graph is $\boxed{\text{decreasing}}$.
   b. $f(x) = 2(1.3^x)$ is an increasing exponential function because $b > 1$ (where $a > 0$ and b are the constants of the exponential model). So $\boxed{\text{the blue graph corresponds to } f}$.
      $g(x) = 2(0.7^x)$ is a decreasing exponential function because $b < 1$ (where $a > 0$ and b are the constants of the exponential model). So $\boxed{\text{the black graph corresponds to } g}$.

c. For $f(x)=2(1.3^x)$ the constant percentage change is
$(b-1)\cdot 100\% = (1.3-1)\cdot 100\%$
$= \boxed{30\%}$

For $g(x)=2(0.7^x)$ the constant percentage change is
$(b-1)\cdot 100\% = (0.7-1)\cdot 100\%$
$= \boxed{-30\%}$

3. a. Both graphs are $\boxed{\text{decreasing}}$.
   b. $f(x)=3(0.7^x)$ decreases less quickly than $g(x)=3(0.8^x)$ because $0.7 < 0.8$. So $\boxed{\text{the blue graph corresponds to } f \text{ and the black graph corresponds to } g}$.
   c. For $f(x)=3(0.7^x)$ the constant percentage change is
   $(b-1)\cdot 100\% = (0.7-1)\cdot 100\%$
   $= \boxed{-30\%}$

   For $g(x)=3(0.8^x)$ the constant percentage change is
   $(b-1)\cdot 100\% = (0.8-1)\cdot 100\%$
   $= \boxed{-20\%}$

5. a. $b > 1$ indicates $\boxed{\text{growth}}$ (where $a > 0$ and b are the constants of the exponential model)
   b. constant percentage change $= (b-1)\cdot 100\%$
   $= (1.05-1)\cdot 100\%$
   $= \boxed{5\%}$

7. a. $b < 1$ indicates $\boxed{\text{decay}}$ (where $a > 0$ and b are the constants of the exponential model)
   b. constant percentage change $= (b-1)\cdot 100\%$
   $= (0.87-1)\cdot 100\%$
   $= \boxed{-13\%}$

9. a. $15 - 3 = \boxed{12 \text{ women}}$
   b. $\dfrac{15-3}{3} = 4 \quad \rightarrow \quad \boxed{400\%}$

11. a. $(1.24-1) = 0.24 \quad \rightarrow \quad \boxed{24\%}$
    b. The number of children born with a genetic birth defect (out of 1000 births) increases by 24% with each additional year of the mother's age (as the mother's age increases from 25 – 49 years).

13. a. $(0.957-1) = -0.043 \quad \rightarrow \quad \boxed{-4.3\%}$
    b. The number of hours of sleep (in excess of 8 hours) that a woman gets each night decreases by 4.3% each year between the ages of 15 and 65.

15. a. $P(t) = 4.81(1.0547^t)$ quadrillion Btu gives the projected amount of petroleum imports t years since 2005, data from $0 \leq t \leq 15$.

b. Solving $P(t) = 10$ yields

$t \approx 13.743 \quad \rightarrow \quad \boxed{2019}$

c. $\boxed{P(t) \text{ increases without bound as } t \text{ increases without bound.}}$

Limit notation for the end behavior is $\boxed{\lim_{t \to \infty} P(t) = \infty}$

17. a. $W(x) = 3.3(0.9854^x)$ gives the number of workers per Social Security beneficiary where x is the number of years since 1996, information from $0 \leq x \leq 34$.

b. $W(34) \approx \boxed{2.001 \text{ workers}}$

19. a. $M(t) \approx 22.242(1.160^t)$ billion dollars gives the projected spending on online marketing where t is the number of years since 2008, data from $0 \leq t \leq 6$.

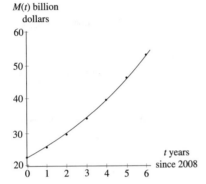

A graph of the model along with a scatter plot of the data shows that the model underestimates the first data point and overestimates the second through the fourth data points.

b. $S(t) \approx 8.180(1.313^t) + 15$ billion dollars gives the projected spending on online marketing where t is the number of years since 2008, $0 \leq t \leq 14$

c.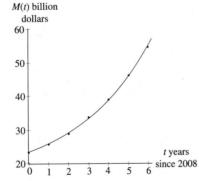

Aligning the data resulted in a better fit for the first six data points. The end behavior is a little higher than the last projected data point.

21. a. (The answer key in the book reports the model and results in this activity to different accuracy levels than reported here. The answer key in the book erroneously identifies the second answer to part c as an answer to part d.)

$f(x) \approx 10.420(0.885^x)$ percent of females aged $17 + x$ years are MySpace users, data from $0 \leq x \leq 18$.

b. $(0.885 - 1) \approx -0.115 \rightarrow \boxed{-11.5\%}$

c. The age 18 years is represented by the input value $x = 1$.

$f(1) \approx 9.222 \rightarrow \boxed{9.2\%}$ (Rounded to reflect the precision of the data.)

The age 20 years is represented by the input value $x = 3$.

$f(3) \approx 7.223 \rightarrow \boxed{7.2\%}$ (Rounded to reflect the precision of the data.)

Both answers are calculated using $\boxed{\text{interpolation}}$.

23. a. A scatter-plot of the data suggests a concave up function so an exponential model is more appropriate than a linear model.

b. linear: $L(x) \approx 1.158x - 50.518$ chirps gives the number of chirps in 13 seconds at a temperature of $x°F$, data from $57 \leq x \leq 80$.

exponential: $E(x) \approx 1.697(1.042^x)$ chirps gives the number of chirps in 13 seconds at a temperature of $x°F$, data from $57 \leq x \leq 80$.

c. Both the linear and exponential models fit the data between $57°F$ and $80°F$ fairly closely. The linear model has the advantage of being simpler to use without the aid of technology than the exponential model.

25. a. $m(x) \approx 0.0023(1.414^x)$ million transistors gives the number of transistors on a computer chip where x is the number of years since 1971 (using Moore's prediction of doubling the number of transistors every two years).

b. $t(x) \approx 0.002(1.425^x)$ million transistors gives the number of transistors on a computer chip where x is the number of years since 1971, data from $0 \leq x \leq 39$.

c. The model from the data is very close to Moore's prediction. In fact, the model from the data shows that the number of transistors doubles a little faster than every two years.

27. a. An exponential function giving the amount of cesium chloride remaining in the body x months after 800 milligrams is injected is $C(x) = 800b^x$ mg.

When $x = 4$ months, there will be only 400 mg of cesium chloride remaining: $C(4) = 400$

Solving $400 = 800b^4$ yields $b \approx 0.841$

$\boxed{C(x) \approx 800(0.841^x) \text{ mg is the amount of cesium chloride remaining in the body } x \text{ months after the injection of 800 mg.}}$

b. 5% of 800 mg is $0.05 \cdot 800 = 40$ mg

Solving $40 = 800(0.841^x)$ yields $x \approx 17.288$ months $\rightarrow \boxed{1 \text{ year, 5 months, 9 days}}$

29. a. An exponential function giving the percentage of the original amount of radon gas that is present after x hours is $r(x) = 100b^x$ percent.

   When $x = 30$ hours, there will be only 80% of the original amount present: $r(30) = 80$

   Solving $80 = 100b^{30}$ yields $b \approx 0.993$

   $\boxed{r(x) \approx 100(0.993^x) \text{ percent of the original amount of radon gas is present after } x \text{ hours.}}$

   b. Solving $50 = 100(0.993^x)$ yields $x \approx 93.189$ hours $\rightarrow$ $\boxed{3 \text{ days, 21 hours, 11 minutes}}$

Section 1.6 Models in Finance (page 64)

1. $1500 + 1500(0.07)(2) = \boxed{\$1710}$

3. a. $F(t) = 15000\left(1 + \dfrac{0.0415}{12}\right)^{12t}$ dollars gives the amount owed after t years.

   b. $\boxed{4.15\%}$

   c. $\left(1 + \dfrac{0.0415}{12}\right)^{12} - 1 \approx 0.0423$ $\rightarrow$ $\boxed{4.23\%}$ (Rounded to match the precision given for APR.)

5. a. $\boxed{12\%}$

   b. $\left(1 + \dfrac{0.12}{12}\right)^{12} - 1 \approx 0.12683$ $\rightarrow$ $\boxed{12.683\%}$

7. a. Solving $2 = 1\left(1 + \dfrac{0.063}{12}\right)^{12t}$ yields $t \approx 11.031$ $\rightarrow$ $\boxed{11 \text{ years, 1 month}}$

   b. Solving $2 = 1 \cdot e^{0.08t}$ yields $t \approx \boxed{8.664 \text{ years}}$

9. a. Option A: For 4.725% compounded semiannually, APY is
   $\left(1 + \dfrac{0.04725}{2}\right)^2 - 1 \approx 0.04781$ $\rightarrow$ $\boxed{4.781\%}$
   Option B: For 4.675% compounded continuously, APY is
   $e^{0.04675} - 1 \approx 0.04786$ $\rightarrow$ $\boxed{4.786\%}$
   Option B has the higher APY so is the better option.

   b. Option A: $1000 deposited for 2 years yields
   $1000\left(1 + \dfrac{0.04725}{2}\right)^{2 \cdot 2} \approx \boxed{\$1097.90}$
   $1000 deposited for 5 years yields
   $1000\left(1 + \dfrac{0.04725}{2}\right)^{2 \cdot 5} \approx \boxed{\$1263.02}$

Option B: $1000 deposited for 2 years yields
$$1000e^{0.04675 \cdot 2} \approx \boxed{\$1098.01}$$
$1000 deposited for 5 years yields
$$1000e^{0.04675 \cdot 5} \approx \boxed{\$1263.33}$$
$\boxed{\text{The length of time deposited does not change the better option.}}$

11. Solving $250{,}000 = P\left(1 + \dfrac{0.04}{12}\right)^{12 \cdot 40}$ yields $P \approx \boxed{\$50{,}608.61}$

13. Solving $56{,}076{,}000{,}000 = 56{,}000{,}000{,}000(1.062)^x$ yields
$x \approx 0.0225$ years $\rightarrow$ $\boxed{\text{less than 2 weeks}}$

Section 1.7

1. a. t years

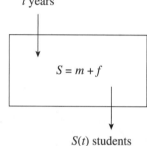

   $S(t)$ students

   b. $S(t) = m(t) + f(t)$ students gives the total number of students in business calculus classes in year t.

3. $\boxed{\text{Functions } f \text{ and } d \text{ cannot be combined into a function giving average credit card debt.}}$ In order to construction such a function, a function giving a monetary amount of credit card debt would be necessary.

5. a. $5.3 - 4.2 = \boxed{\$1.1 \text{ million}}$
   b. $P(q) = R(q) - C(q)$ million dollars gives the profit from the production and sale of q units of a commodity.

7. a. $72 + 129 = \boxed{201 \text{ billion Euros}}$
   b. $R(t) = C(t) + P(t)$ billion Euros gives the profit for a company during the tth quarter.

9. a. $5 - (-3) = \boxed{\$8 \text{ billion}}$
   b. $C(t) = R(t) - P(t)$ billion dollars gives the cost during the tth quarter.

11. a. $N(t) = E(t) - 1000 \cdot I(t)$ trillion Btu gives the net trade of natural gas in year t.
    b. A negative net trade value indicates that the value of imports exceeded the value of exports.

1.7 Constructed Functions
Solutions to Odd-Numbered Activities

13. a. $\overline{R}(t)\dfrac{\text{dollars}}{\text{gallon}} \cdot P(t)\dfrac{2010 \text{ dollars}}{\text{dollars (at pump)}} = (\overline{R} \cdot P)(t)\dfrac{2010 \text{ dollars}}{\text{gallon}}$

    b. 2010-dollars per gallon

15. a. $\overline{A}(t) = \dfrac{D(t)}{N(t)}$ thousand dollars per cardholder

    b. $\overline{A}(t)$ thousand dollars gives the average amount of credit card debt per cardholder in year t.

17. a. $\boxed{(f+g)(x) = 5x + 4 + 2x^2 + 7}$

    Alternate form: $(f+g)(x) = 2x^2 + 5x + 11$

    $(f+g)(2) = \boxed{29}$

    b. $\boxed{(f-g)(x) = 5x + 4 - (2x^2 + 7)}$

    Alternate form: $(f-g)(x) = -2x^2 + 5x - 3$

    $(f-g)(2) = \boxed{-1}$

    c. $\boxed{(f \cdot g)(x) = (5x+4) \cdot (2x^2 + 7)}$

    Alternate form: $(f \cdot g)(x) = 10x^3 + 8x^2 + 35x + 28$

    $(f \cdot g)(2) = \boxed{210}$

    d. $\boxed{\dfrac{f}{g}(x) = \dfrac{5x+4}{2x^2+7}}$

    $\dfrac{f}{g}(2) = \dfrac{14}{15} \approx \boxed{0.933}$

19. a. $\boxed{(p+s)(t) = -2t^2 + 6t - 4 + 5t^2 + 2t + 7}$

    Alternate form: $(p+s)(t) = 3t^2 + 4t + 3$

    $(p+s)(2) = \boxed{23}$

    b. $\boxed{(p-s)(t) = -2t^2 + 6t - 4 - (5t^2 + 2t + 7)}$

    Alternate form: $(p-s)(t) = -7t^2 + 8t - 11$

    $(p-s)(2) = \boxed{-23}$

    c. $\boxed{(p \cdot s)(t) = (-2t^2 + 6t - 4) \cdot (5t^2 - 2t + 7)}$

    Alternate form: $(p \cdot s)(t) = -10t^4 + 34t^3 - 46t^2 + 50t - 28$

    $(p \cdot s)(2) = \boxed{0}$

    d. $\boxed{\dfrac{p}{s}(t) = \dfrac{-2t^2 + 6t - 4}{5t^2 - 2t + 7}}$

    $\dfrac{p}{s}(2) = \boxed{0}$

21. a.   *t* hours

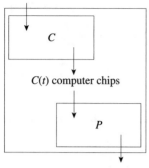

P(C(t)) dollars

b. $P \circ C(t)$ dollars is the profit generated from the sale of computer chips produced after *t* hours

23. a.   *t* hours

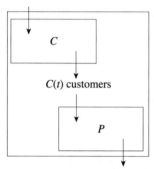

P(C(t)) dollars

b. $P \circ C(t)$ dollars is the average amount of tips generated *t* hours after 4 P.M.

25. $\boxed{f(t(p)) = 3e^{4p^2}}$

$f(t(2)) = f(16) \approx \boxed{26{,}658{,}331.562}$

27. $\boxed{g(x(w)) = \sqrt{7(4e^w)^2}}$

Alternate form: $g(x(w)) = 4e^w \sqrt{7}$

$g(x(2)) = g(29.556) = \boxed{78.198}$

29. *y* is a function because a vertical line intersects the graph in at most one point. *y* is one-to-one because any horizontal line intersects the graph in at most one point.

31. *y* is a function because a vertical line intersects the graph in at most one point. *y* is not one-to-one because there are horizontal lines that intersect the graph in as many as three points.

33. *y* is not a function because there is a vertical line intersects the graph in two points.

35. $P(d) = 0.030d + 1$ atm gives the pressure at a depth of *d* feet under water, data from $0 \leq d \leq 32$.

$D(p) = -33p + 33$ feet gives the depth under water where the pressure is *p* atm, data from $1 \leq p \leq 5$.

37. $D(g) \approx 4.114g - 0.001$ dollars gives the price of g gallons of gas, data from $0 \leq g \leq 20$.
$G(d) \approx 0.243d$ gallons gives the amount of gas that can be purchased for d dollars, data from $0 \leq d \leq 82.28$.

39. $f(x) = 5x + 7$

  $y = 5x + 7$ function output rewritten as y

  $y - 7 = 5x$ 7 subtracted from both sides

  $\dfrac{y-7}{5} = x$ both sides divided by 5

  $\boxed{x(f) = \dfrac{f-7}{5}}$ variables rewritten

41. $g(t) = \ln t$

  $y = \ln t$ function output rewritten as y

  $e^y = e^{\ln t}$ set both sides as powers of e

  $e^y = t$ rewrite $e^{\ln t}$ as t

  $\boxed{t(g) = e^g}$ variables rewritten

43. $y(x) = 73x - 35.1$

  $y = 73x - 35.1$ function output rewritten as y

  $y + 35.1 = 73x$ add 35.1 to both sides

  $\dfrac{y + 35.1}{73} = x$ divide both sides by 73

  $\boxed{x(y) = \dfrac{y + 35.1}{73}}$ variables rewritten

45. $h(t) = 0.5 + \dfrac{5}{0.2t}$

  $y = 0.5 + \dfrac{5}{0.2t}$ function output rewritten as y

  $y - 0.5 = \dfrac{5}{0.2t}$ subtract 0.5 to both sides

  $\dfrac{0.2}{5}(y - 0.5) = \dfrac{1}{t}$ multiply both sides by $\dfrac{0.2}{5}$

  $t = \dfrac{5}{0.2(y - 0.5)}$ take the reciprocal of each side

  $\boxed{t(h) = \dfrac{5}{0.2(h - 0.5)}}$ variables rewritten

  Alternate form: $t(h) = \dfrac{25}{(h - 0.5)}$

Section 1.8 Logarithmic Functions and Models (pages 81–86)

1. a. The black graph is decreasing.
   The blue graph is increasing.
   b. The black graph is concave up.
   The blue graph is concave down.
   c. $f(x) = 1 + 2\ln x$ is an increasing, concave down natural log function because $b > 0$ (where a and b are the constants of the natural log function). So the blue graph corresponds to f.
   $g(x) = 3 - 2\ln x$ is a decreasing, concave up natural log function because $b < 0$ (where a and b are the constants of the natural log function). So the black graph corresponds to g.

3. a. Both graphs are decreasing.
   b. Both graphs are concave up.
   c. $f(x) = 3.4 - 7\ln x$ is stretched more vertically than $g(x) = 3 - 2\ln x$ because the scale factor for f, 7, is greater than the scale factor for g, 2.
   So the black graph corresponds to f and the blue graph corresponds to g.

5. a. decreasing, concave up
   b. Both an exponential function with percentage change $b < 1$ or a logarithmic function with constant multiplier $b < 0$ are decreasing and concave up.

7. a. increasing, concave up.
   b. A positive exponential function with percentage change $b > 1$ is increasing and concave up. This scatter plot shows no evidence of a vertical asymptote to the left which is a characteristic of logarithmic functions.

9. a. concave down
   b. There is no evidence that the output will level out or decrease as input increases, so if the output continues to follow the pattern shown by the scatter plot it will increase without bound as the input values increase without bound: $\lim_{t \to \infty} C(t) = \infty$
   c. The differences between output values for consecutive input values are larger for smaller input values than for larger input values.
   d. $C(t) \approx -1{,}585.229 + 1{,}435.942 \ln t$ billion dollars, where t is the number of years since 1990, data from $10 \le t \le 18$.

11. a. $C(x) \approx 1.182 + 2.216 \ln x$ µg/ml gives the concentration of piroxicam in the bloodstream after x days of taking the drug, data from $1 \le x \le 17$.
    b. $\lim_{x \to \infty} C(x) = \infty$
       $\lim_{x \to 0^+} C(x) = -\infty$
    c. (The answer key in the book reports this result to less accuracy.)
       $C(2) \approx 2.718$ µg/ml

13. a.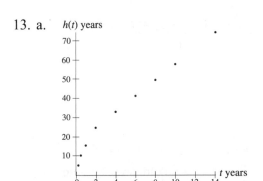

The scatter plot suggests an increasing, concave down function.

b. $h(t) \approx -20.210 + 30.932 \ln t$ years gives the human equivalent for a dog aged $t-2$ years, data for dogs aged 3 months to 14 years.

15. a. $s(34) \approx$ 66.519 million subscribers
$s(38) \approx$ 65.798 million subscribers
$s(42) \approx$ 65.148 million subscribers

b. Interchanging the input and output values from part *a* results in the following table:

s million subscribers	r dollars per month
66.519	34
65.798	38
65.148	42

$r(s) \approx 967{,}186.056(0.857^s)$ dollars gives the cable rate at which there will be s million subscribers, calculated over $34 \leq r \leq 42$. (The answer key in the book incorrectly states this interval as being an interval over s.)

17. a. $L(x) \approx 158.574 - 42.877 \ln x$ ppm gives the concentration of lead in the soil x meters from a heavily traveled road, data from $5 \leq x \leq 20$.

b. $L(12) \approx 52.029$ → 52 ppm (Rounded to nearest integer.)

c. $E(x) \approx 123.278(0.932^x)$ ppm gives the concentration of lead in the soil x meters from a heavily traveled road, data from $5 \leq x \leq 20$.
An exponential model better displays the end behavior of the context than does a log model. The end behavior of the exponential model indicates that the lead concentration will decrease toward zero as the distance from the road increases. The end behavior of the log model indicates that lead concentrations will eventually become negative as distance increases. Negative lead concentration does not make sense in context.

19. $D(c) \approx 0.603(1.562^c)$ days gives the length of time needed to achieve a concentration of c μg/ml of piroxicam in the bloodstream where d is the number of days that the drug has been taken, data from $1.5 \leq c \leq 7.5$.

21. $t(h) \approx 1.966(1.032^h)$ years gives the equivalent dog's age for a human aged h years, data from $5 \leq h \leq 91$.

23. $A(x) \approx 3.495 + 10.175 \ln x$ years gives the age of a bluefish of length x inches, data from $18 \leq x \leq 32$.

25. a. $s(15) \approx 1.395$ hours → 9 hours, 24 minutes

$s(20) \approx 1.120$ hours → 9 hours, 7 minutes

$s(40) \approx 0.465$ hours → 8 hours, 28 minutes

$s(64) \approx 0.162$ hours → 8 hours, 10 minutes

b. $w(s) \approx 22.563 - 22.767 \ln s$ years gives the age of a woman who gets $s + 8$ hours of sleep per night, information from $0.162 \le s \le 1.393$. (The answer key in the book incorrectly states the function with the natural log of x instead of the natural log of s.)

27. Answers will vary but should include the following information:
Decreasing exponential functions and decreasing log functions are both concave up. However, an exponential function approaches a horizontal asymptote as input values increase without bound, while a log function decreases without bound as input values increase without bound. When the context of a data set implies that the output values of the function approach and remain near a specific value as input increases, an exponential model is appropriate.

Section 1.9 Quadratic Functions and Models (pages 91–93)

1. a. concave up
   b. minimum occurs at $x \approx 3$
   c. decreasing over $-0.5 < x < 3$
      increasing over $3 < x < 4$

3. a. concave down
   b. maximum occurs at $t \approx 2$
   c. increasing over $0 < t < 2$
      decreasing over $2 < t < 4.35$ (The answer key in the book incorrectly states that the function is increasing over $2 < t < 4$ and ignores the portion of the graph to the right of $t = 4$.)

5. a. concave up
   b. no maximum or minimum shown
   c. increasing for all t shown

7. (The answer key in the book incorrectly states that a logarithmic function is suggested by the scatter plot.)
The increasing, concave up scatter plot suggests either an exponential or a quadratic function.

9. The decreasing, concave down scatter plot suggests a quadratic function.

11. (The answer key in the book incorrectly states that an exponential function is suggested by the scatter plot.)
The decreasing, concave down scatter plot becomes negative suggesting a logarithmic or a quadratic function.

13. The concave up scatter plot with an obvious minimum suggests a quadratic function.

15. a.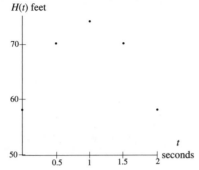

A scatter plot of the data indicates a maximum point and changes from increasing to decreasing at that point. Logarithmic and exponential functions do not possess maximum points but quadratic functions do.

b. $H(t) = -16t^2 + 32t + 58$ feet gives the height of a rocket above the surface of a pond t seconds after the rocket was launched, data from $0 \le t \le 2$.

c. Solving $H(t) = 0$ yields $x \approx -1.151$ (The negative solution is not valid in this context.) and $x \approx 3.151$ → $\boxed{3.15 \text{ seconds}}$ (Rounded to the nearest hundredth.)

17. a.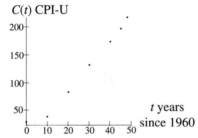

The increasing, concave up scatter plot of this data suggests a quadratic or an $\boxed{\text{exponential}}$ function.

b. The $\boxed{\text{quadratic}}$ model seems to follow the pattern of the data better than the exponential model which overestimates the 2005 and 2008 data.

$\boxed{C(t) \approx 0.038t^2 + 2.143t + 24.089 \text{ gives the CPI (based on a CPI} = 100 \text{ for 1982-84) for all U.S. urban consumers where } t \text{ is the number of years since 1960, data from } 0 \le t \le 48.}$

19. a.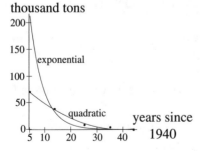

The quadratic model appears to pass through the first and last data points and passes relatively close to each of the other three points. The exponential model passes relatively close to the second and last data points, but is farther away from the other points and appears to have a much more accentuated curvature than the scatter plot of the data suggests. The quadratic model appears to have the better fit.

b.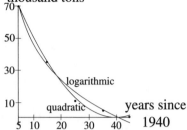

(The answer key in the book incorrectly chooses the logarithmic model over the quadratic model even though it does not fit better.)
The logarithmic model appears to pass through the first and fourth points. It passes relatively close to the second point but misses the third point and is negative beginning in 1975. Because the context does not allow for negative output, the logarithmic model cannot be considered as a valid model. The quadratic model is better.

c. The banning of lead in paint in 1976 suggests that an exponential function would fit the asymptotic end behavior better than a quadratic function.

d. $u(x) \approx 0.057x^2 - 4.558x + 90.789$ thousand tons gives the amount of lead used in paint x years since 1935, data from $5 \le x \le 45$.

Section 1.10 Logistic Functions and Models (pages 98–102)

1. a. $x \approx 3$
   b. increasing over all x shown
      concave up: $-10 < x < 3$
      concave down: $3 < x < 10$

3. a. $x \approx 0$
   b. decreasing over all x shown
      concave down: $-4 < x < 0$
      concave up: $0 < x < 4$

(The instructions for Activities 5 through 10 on page 98 of the text should include *logistic* functions as one of the possible choices in part b.)

5. a. no The scatter plot suggests only one concavity – concave down.
   b. The increasing, concave down scatter plot suggests either a log or a quadratic function.

7. a. yes The scatter plot suggests a change from concave up to concave down between the fourth and fifth points.
   b. The increasing scatter plot with its change from concave up to concave down suggests a logistic function.

9. a. no The scatter plot suggests no curvature; and therefore, no change in concavity.
   b. The scatter plot with no curvature suggests a linear function.

1.10 Logistic Functions and Models
Solutions to Odd Numbered Activities

11. a. The presence of a negative sign in the exponent on e in the denominator signifies an boxed{increasing} logistic function.
    b. $\lim_{x \to -\infty} f(x) = 0$
       $\lim_{x \to \infty} f(x) = 100$
    c. lower asymptote: $y = 0$
       upper asymptote: $y = 100$

13. a. The absence of a negative sign in the exponent on e in the denominator signifies an boxed{decreasing} logistic function.
    b. $\lim_{t \to -\infty} s(t) = 10.2$
       $\lim_{t \to \infty} s(t) = 0$
    c. upper asymptote: $y = 10.2$
       lower asymptote: $y = 0$

15. a. The scatter plot appears to be decreasing toward a horizontal asymptote at $y = 0$. The context indicates that there should be an upper asymptote at $y = 100$.
    b. $f(x) \approx \dfrac{103.880}{1 + 0.167 e^{0.510x}}$ percent gives the percentage of residential Internet that was accessed by dial-up x years since 2000, data from $0 \le x \le 14$. (Overflow occurs if input values are not aligned.)
    c. upper asymptote: boxed{$y \approx 103.880$}
       lower asymptote: boxed{$y = 0$}

17. a. $f(t) \approx \dfrac{251.299}{1 + 0.138 e^{0.285t}}$ million tons gives the amount of lead emissions into the atmosphere t years since 1970, data from $0 \le t \le 25$. (Overflow occurs if input values are not aligned.)
    b. million tons

    [Graph showing a decreasing logistic curve with y-axis values 500, 100, 150, 200, x-axis "years since 1970" from 0 to 25, with inflection point marked]

    $t \approx 7 \rightarrow$ boxed{1977}
    (The answer key in the text erroneously reports the inflection point as occurring in 1978.)
    c. concave down: $t < 7$ before 1977
       concave up: $t > 7$ after 1977

19. (The equation and results from the equation are incorrect in the answer key in the text.)
    a. $f(w) \approx \dfrac{3015.991}{1 + 331.884 e^{-1.024w}}$ deaths gives the number of deaths that had occurred among U.S. Navy personnel by the end of the wth week after 8/24/1918, data from $1 \le w \le 14$.

b. $3137 - 3015.991 \approx 121.009 \rightarrow \boxed{121 \text{ deaths}}$

21. a. $f(w) = \dfrac{91{,}317.712}{1 + 3057.699e^{-0.951w}}$ deaths gives the number of deaths that had occurred among civilians by the end of the wth week after 8/24/1918, data from $3 \le w \le 14$.

b. Answers will vary but might include:
1) the spread of information about the influenza and how to avoid contracting it slows down the spread of influenza,
2) the onset of colder weather in November leads to more isolation of citizens and retards the spread of diseases,
3) the spread of disease is faster through certain dense subsections of a city and then slows down as those subsections reach saturation levels.

23. a. $s(t) \approx \dfrac{249.969}{1 + 91.546e^{-0.617t}}$ bases gives the cumulative number of bases stolen by Willie Mays t years since 1950, data from $1 \le t \le 13$. (Overflow occurs if input values are not aligned.)

b. $s(14) - s(13) \approx \boxed{3 \text{ bases}}$

c. The model $\boxed{\text{underestimates by 16 bases}}$.

25. a. $m(t) \approx \dfrac{1390.487}{1 + 67.888e^{-0.267t}} + 4000$ million metric tons gives the amount of carbon dioxide emissions released into the atmosphere during the tth year since 1980, data from $0 \le t \le 24$. (Domain error occurs with no output alignment. Overflow occurs when output is aligned but input is not.)

b. Subtracting 4000 from each output value shifts the points on the scatter plot down so that they appear to approach 0 as input approaches 1980 from the right.
Adding 4000 to a model of the shifted data shifts the function up so that the model
$m(t) =$ shifted model $+ 4000$
fits the un-aligned data.

27. Answers will vary but should be similar to the following:
The graph of an increasing logistic function begins near zero. It is increasing, concave up until it reaches an inflection point. It continues to increase but is concave down and approaches a horizontal asymptote.
The graph of a decreasing logistic function begins near an upper horizontal asymptote. It is decreasing, concave down until it reaches an inflection point. It continues to decrease but is concave up and approaches the input axis asymptotically.

Section 1.11 Cubic Functions and Models (pages 107–110)

1. a. $x \approx 10$

b. increasing: $0 < x < 4$ and $16.25 < x < 20$
decreasing: $4 < x < 16.25$
concave down: $0 < x < 10$
concave up: $10 < x < 20$

3. a. $x \approx 390$
    b. increasing: $0 < x < 230$
       decreasing: $230 < x < 500$
       concave down: $0 < x < 390$
       concave up: $390 < x < 500$

5. a. $x \approx 74$
    b. decreasing: $0 < x < 14$ and $134 < x < 140$
       increasing: $14 < x < 134$
       concave up: $0 < x < 74$
       concave down: $74 < x < 140$

7. a. [no] The scatter plot suggests only one concavity – concave down.
    b. The increasing, concave down scatter plot suggests a [logarithmic] function because output values appear to increase more slowly as input values increase but there is no indication that output values will level out or reach a maximum and then decrease.

9. a. [yes] The scatter plot suggests a change from concave up to concave down between the fourth and fifth points.
    b. The decreasing scatter plot with its change in curvature from concave up to concave down suggests a [cubic] function.

11. a. [yes] The scatter plot suggests a change from concave down to concave up near the fourth point.
    b. The scatter plot changes concavity from concave down to concave up and also appears to have a minimum suggesting a [cubic] function.

13. a. [no] The scatter plot suggests only one concavity – concave up.
    b. The decreasing, concave up scatter plot becomes negative after the fourth point suggesting a [logarithmic] function or possibly [quadratic] function.

15. a. $f(x) = x^3 - 5.143x^2 + 1.571x + 59.914$ billion dollars gives the monetary value of loss resulting from identity fraud x years since 2004, data from $0 \le x \le 4$.
    b. $f(5) \approx$ [$64.2 billion]
       There is no evidence to indicate that loss due to identity theft continued according to this cubic trend beyond the data given.
    c. The only output values with valid interpretation in context correspond to integer input values because the function gives yearly totals.

17. a.
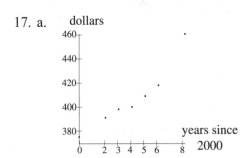

A scatter plot of the data suggests a cubic function because it is increasing and is concave down before 2004 and concave up after 2004.

b. $s(t) \approx 338t^3 - 3.015t^2 + 13.223t + 374.860$ dollars gives the median weekly salary of 16-24 year old men employed full-time t years since 2000, data from $0 \le t \le 8$.

c. $s(10) \approx 543.860 \rightarrow$ $\boxed{\$544}$

d. 2010 is outside the data interval so the result to part c is found using $\boxed{\text{extrapolation}}$.

19. a. trillion dollars

    [scatter plot with axes: trillion dollars (vertical, 0 to 7) vs million SUVs (horizontal, 20 to 60)]

    A scatter plot of the data is $\boxed{\text{increasing, concave down}}$ suggesting either a $\boxed{\text{logarithmic}}$ or a $\boxed{\text{quadratic}}$ function.

    b. Profit from SUV sales will most likely $\boxed{\text{increase without bound}}$ as more SUVs are sold (assuming new models are being produced to replace older models). This continued increase suggests a $\boxed{\text{logarithmic}}$ model instead of a quadratic model which would reach a maximum and then decrease.

    c. $P(q) \approx -6.112 + 3.059 \ln q$ trillion dollars gives the profit from the production and sale of q million SUVs, data from $10 \le x \le 70$.

    d. same as part c

21. a. billion units

    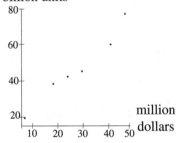

    A scatter plot of the data is increasing and appears to have an inflection point near $x = 24$. The scatter plot is concave down to the left of $x \approx 24$ and concave up to the right of $x \approx 24$. This behavior suggests a $\boxed{\text{cubic}}$ function.

    b. Production should continue to increase as capital expenditure increases (assuming appropriate levels of labor are also maintained). However, it is unlikely production will continue to follow a cubic increase indefinitely.
    cubic (for close extrapolations only)

    c. $P(x) = 0.002x^3 - 0.149x^2 + 4.243x - 1.550$ billion units gives the production level when x million dollars is invested in capital, data from $6 \le x \le 48$. (Rounding coefficients significantly changes the results calculated from this model.)

    d. same as c (for close extrapolations only)

23. a.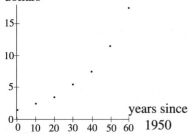

A scatter plot of the data is increasing, concave up. This behavior suggests either a ⬚quadratic⬚ or an ⬚exponential⬚ function.

b. It makes sense to assume that price will continue to increase as time increases and that the increase will follow the same inflation trends as in the past.
quadratic or exponential

c. $f(x) \approx 1.582(1.041^x)$ dollars gives the retail price for souvenir footballs at CU x years since 1950, data from $0 \leq x \leq 60$. (Rounding coefficients significantly changes the results calculated from this model.)

d. same as part c

25. Answers will vary in form but should contain the same elements as follows:
Functions of the form $f(x) = ax^2 + bx + c$ are referred to as quadratic functions and have graphs that are either concave down for all x or concave up for all x. Graphs of quadratic functions change direction once: a concave down quadratic increases from the left and decreases to the right, and a concave up quadratic decreases from the left and increases to the right.
Functions of the form $g(x) = ax^3 + bx^2 + cx + d$ are referred to as cubic functions and have graphs that change concavity exactly once. Graphs of cubic functions either change direction twice (from increasing to decreasing and back to increasing or from decreasing to increasing and back to decreasing) or they never change direction. A cubic function that is increasing from the left is concave down to the left of the inflection point and concave up to the right of it. A cubic function that is decreasing from the left is concave up to the left of the inflection point and concave down to the right.

Section 1.12 Cyclic Functions and Models (pages 115–119)

1.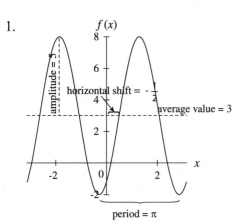

3.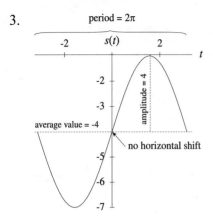

5. amplitude: $a = \boxed{1}$
   period: $\dfrac{2\pi}{b} = \dfrac{2\pi}{2.2} \approx \boxed{2.856}$
   average value: $d = \boxed{0.7}$
   horizontal shift: $\dfrac{c}{b} = \dfrac{0.4}{2.2} \approx \boxed{0.182}$

7. amplitude: $a = \boxed{3.62}$
   period: $\dfrac{2\pi}{b} = \dfrac{2\pi}{0.22} \approx \boxed{28.560}$
   average value: $d = \boxed{7.32}$
   horizontal shift: $\dfrac{c}{b} = \dfrac{4.81}{0.22} \approx \boxed{21.864}$

9. amplitude: $a = \boxed{1}$
   period: $\dfrac{2\pi}{b} = \dfrac{2\pi}{\pi} \approx \boxed{2}$
   average value: $d = \boxed{0}$
   horizontal shift: $\dfrac{c}{b} = \dfrac{-2}{\pi} \approx \boxed{-0.637}$

11. amplitude: $a = \boxed{1}$
    period: $\dfrac{2\pi}{b} = \dfrac{2\pi}{1} \approx \boxed{6.283}$
    average value: $d = \boxed{0}$
    horizontal shift: $\dfrac{c}{b} = \dfrac{0}{1} = \boxed{0}$

13.

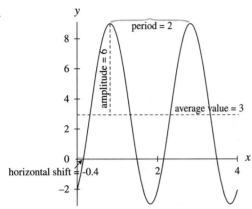

a. amplitude: $\dfrac{\text{max}-\text{min}}{2} = \dfrac{9-(-3)}{2} = \boxed{6}$

period: $\boxed{2}$ (see figure)

average value: $\dfrac{\text{max}+\text{min}}{2} = \dfrac{9+(-3)}{2} = \boxed{3}$

horizontal shift: $\boxed{-0.4}$ (see figure)

b. Solving period $\dfrac{2\pi}{b} = 2$ for b gives $b = \pi$.

Solving horizontal shift $\dfrac{c}{\pi} = -0.4$ for c gives $c = -0.4\pi$.

$\boxed{f(x) = 6\sin(\pi x - 0.4\pi) + 3}$

15.

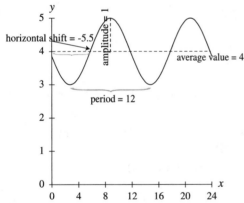

a. amplitude: $\dfrac{\text{max}-\text{min}}{2} = \dfrac{5-3}{2} = \boxed{1}$

period: $\boxed{12}$ (see figure)

average value: $\dfrac{\text{max}+\text{min}}{2} = \dfrac{5+3}{2} = \boxed{4}$

horizontal shift: $\boxed{-5.5}$ (see figure)

b. Solving period $\frac{2\pi}{b}=12$ for b gives $b=\frac{2\pi}{12}\approx 0.524$.

Solving horizontal shift $\frac{c}{\frac{\pi}{6}}=-5.5$ for c gives $c=(-5.5)\frac{\pi}{6}\approx -2.880$.

$$\boxed{f(x)\approx \sin(0.524x-2.880)+4}$$

17. a. amplitude: $a=\boxed{37°F}$

average value: $d=\boxed{25°F}$

b. high: $d+a=25+37=\boxed{62°F}$

low: $d-a=25-37=\boxed{-12°F}$

The mean daily temperature in Fairbanks can be as high as 62°F and as low as −12°F.

c. period: $\frac{2\pi}{0.0172}\approx \boxed{365.3 \text{ days}}$

yes; Because the period of the model is one year, the model will continue to fit beyond the first year.

19. a. maximum: $\boxed{24 \text{ hours}}$
minimum: $\boxed{0 \text{ hours}}$

amplitude: $\frac{\text{max}-\text{min}}{2}=\frac{24-0}{2}=\boxed{12 \text{ hours}}$

average value: $\frac{\text{max}+\text{min}}{2}=\frac{24+0}{2}=\boxed{12 \text{ hours}}$

b. period: $\boxed{365 \text{ days}}$

horizontal shift: $\boxed{-81 \text{ days}}$ (midway between winter and summer solstice)

Solving the period formula $\frac{2\pi}{b}=365$ yields $b\approx \boxed{0.017}$

Solving the horizontal shift formula $\frac{c}{\frac{2\pi}{365}}=-81$ yields $c\approx \boxed{-1.394}$

c. $f(x)\approx 12\sin(0.017x-1.394)+12$ hours gives the amount of daylight on the xth day of the year, data from $0\le x\le 365$.

21. a. period: $\boxed{52 \text{ weeks}}$

horizontal shift: $\boxed{-44 \text{ weeks}}$ (midway between week 31 and week 57)

Solving the period formula $\frac{2\pi}{b}=52$ yields $b\approx \boxed{0.121}$

Solving the horizontal shift formula $\frac{c}{\frac{2\pi}{52}}=-44$ yields $c\approx \boxed{-5.317}$

b. amplitude: $\frac{\text{max}-\text{min}}{2}=\frac{215-100}{2}=\boxed{57.5 \text{ million cans}}$

average value: $\frac{\text{max}+\text{min}}{2}=\frac{215+100}{2}=\boxed{157.5 \text{ million cans}}$

c. $s(w)=57.5\sin(0.121w-5.317)+157.5$ million cans gives the number of soup cans sold each week w weeks since the end of the previous year, data from $0\le w\le 52$.

23. a. Mean temperatures rise and fall with the changing seasons so are cyclic in nature.
   b. $t(x) \approx 19.336\sin(0.550x - 2.366) + 73.687$ degrees Fahrenheit gives the normal daily mean temperature for Phoenix in the xth month of the year (i.e., $x = 1$ represents January), data from $0 \le x \le 12$.
   c. period $= \dfrac{2\pi}{b}$
   $\approx \dfrac{2\pi}{0.550}$
   $\approx \boxed{11.4 \text{ months}}$
   The period generated by the model is approximately half a month less than what would be expected from the context.
   d. $t(7) \approx \boxed{93.0°\text{F}}$
   The model gives a very close estimate of normal July temperatures.

25. a. Natural gas is used for heating homes, so natural gas usage will cycle opposite the seasonal changes in temperature.
   b. $g(m) \approx 1.610\sin(0.534m + 0.322) + 1.591$ therms/day gives the average natural gas usage in Reno in the mth month of the year (i.e., $m = 1$ represents January), data from $-1 \le x \le 13$.
   c. period $= \dfrac{2\pi}{b}$
   $\approx \dfrac{2\pi}{0.534}$
   $\approx \boxed{11.8 \text{ months}}$
   The period generated by the model is slightly less than one year. Extrapolations should be relatively accurate for the first year beyond the data. By the fifth year beyond the input interval of the data, the extrapolations will be off by an entire month.

27. a. billion trips

   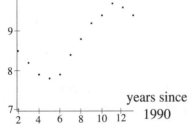

   A scatter plot of the data suggests one change in concavity and two changes in direction. It is concave up to the left of 1997 and concave down to the right of 1997. This behavior suggests either a $\boxed{\text{sine}}$ function or a $\boxed{\text{cubic}}$ function.
   b. no; It is reasonable to assume that mass transit use will continue to decline for a short period after 2003, but there is no evidence that it will continue to decline indefinitely (cubic), nor is there evidence that it will periodically alternate between two extremes (sine).
   c. $t(x) \approx 0.920\sin(0.481x - 2.430) + 8.753$ billion trips gives the number of mass transit trips in the United States x since 1990, data from $2 \le x \le 13$.
   d. Even though both the cubic model and the sine model closely fit the data, the sine model stays closer to more points than does the cubic.

Chapter 1 Review Activities (pages 122–127)

1. a. A formula is an $\boxed{\text{algebraic}}$ representation.
    b. Input is $\boxed{\text{years since 2004}}$ and the units of measure are $\boxed{\text{years}}$.
       Output is $\boxed{\text{the number of phones recycled}}$ and is $\boxed{\text{million phones}}$.
    c. The relation is $\boxed{\text{a function}}$ because for any given year there can be only one corresponding number of phones recycled.
    d.

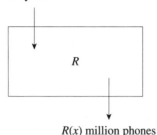

$R(x)$ million phones

3. a. The statement is a $\boxed{\text{verbal}}$ representation.
    b. Input is $\boxed{\text{the number of letters}}$ in a word and the units of measure are $\boxed{\text{letters}}$.
       Output is $\boxed{\text{the number of syllables}}$ in a word and is measured in $\boxed{\text{syllables}}$.
    c, d. The relation is $\boxed{\text{not a function}}$ because two words with the same number of letters may have different number of syllables.

5. a. $I(11) = \boxed{4.27 \text{ million teenagers}}$
       $\boxed{\text{At the end of 2011, 4.27 million American teenagers had internet access.}}$
    b. The result from part a is found using $\boxed{\text{interpolation}}$ because $t = 11$ is within the input interval $9 \le t \le 13$.
    c. Solving $I(t) = 5$ gives $t \approx 12.674 \quad \rightarrow \quad \boxed{\text{August 2013}}$

7. a. $t \approx 4$
    b. increasing: $0 < t < 3$ and $5 < t < 13$
       decreasing: $3 < t < 5$
       concave down: $0 < t < 4$
       concave up: $4 < t < 13$
    c. In general, between 1995 and 2008 the percentage of drug prescriptions allowing generic substitution was increasing. There was a short period between 1998 and 2000 when the percentage dropped just slightly before rising again.

9. a. (The expression for y is set incorrectly in the text, it should read $y(t) = \dfrac{9.795}{1+15.75e^{-0.354t}} + 5$.)

$t \to \infty$	$y(t)$
10	11.723
20	14.667
40	14.795
80	14.795
160	14.795
$\lim\limits_{t \to \infty}[y(t)] \approx 14.8$	

b. $y = 5$
$y = 14.795$

c. In the long run, approximately 14.8 pounds of yogurt will be available for each person in the United States.

11. a. $\lim\limits_{x \to 3^-} f(x) = \boxed{-5.5}$

b. $\lim\limits_{x \to 3^+} f(x) = \boxed{-5.5}$

c. $\lim\limits_{x \to 3} f(x) = \boxed{-5.5}$

d. $f(3) = \boxed{-2.5}$

e. The function f is not continuous at $x = 3$ because the limit does not equal the function value at $x = 3$.

13. a. constant rate of change

b. $f(x) = 2.6x + 93$ million men gives the number of men who will be using the Internet x years after 2008, data from $0 \le x \le 5$.

c. slope: $\boxed{2.6 \text{ million users per year}}$
y-intercept: $\boxed{93 \text{ million users}}$

15. a. constant percentage change

b. $f(m) = 76(1.08^m)$ million people gives the number MySpace members m months after the end of 2008, data from $0 \le m \le 48$.

c. y-intercept: $\boxed{76 \text{ million members}}$
growth rate: $\boxed{0.8\%}$

17. a. constant percentage change

b. $f(x) = 40(0.928^x)$ mg gives the amount of Prozac still in the bloodstream x days after dosage.

c. y-intercept: $\boxed{40 \text{ mg}}$
decay rate: $\boxed{7.2\%}$

19. a. $R(q) = q \cdot p(q)$ dollars gives revenue from the sale of q bottles of shampoo.

b. $P(q) = q \cdot p(q) - C(q)$ dollars gives the profit from the sale of q bottles of shampoo.

c. $\overline{P}(q) = \dfrac{q \cdot p(q) - C(q)}{q}$ dollars per bottle is the average profit from the sale of q bottles of shampoo.

21. a. $t(x) = n(x) + \dfrac{f(x)}{1000}$ million passengers gives the total number of cruise passengers x years since 1991, data from $0 \le x \le 17$.

    b. $p(x) = \dfrac{n(x)}{t(x)} \cdot 100\%$ gives the percentage of cruise passengers that are from North America x years since 1991, data from $0 \le x \le 17$.

23. a.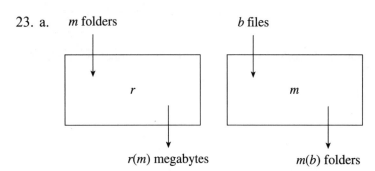

    b. The function $r \circ m$ gives the number of megabytes of memory reserved for the mailbox folders containing b email files.

    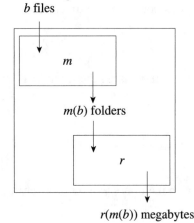

25. a. Approximately 904 lives are saved each year by the enforcement of minimum drinking-age laws.
    b. (The equation in the answer key in the text is incorrect.)
       1997: $s(0) \approx 17.329$
       2007: $s(10) \approx 26.369$
       $t(s) = 1.106s - 19.169$ years gives the number of year since 1997 in which an estimated s thousand lives were saved because of minimum drinking-age laws.

27. a. A scatter plot of the data is concave down. It is increasing to the left of 19 and decreasing to the right of 19. It would be appropriate to use a quadratic function to model this behavior. It would not be appropriate to use a cubic function because there is no evidence of an inflection point.
   b. $f(x) \approx -34.768x^2 + 1363.332x - 12{,}580.25$ drivers gives the number of x-year-old drivers who died in automobile accidents.
   c. $f(19) \approx 772$ drivers
   $790 - 772 = 18$ drivers

29. a. A scatter plot appears to be concave up. It is decreasing to the left of 2002 and increasing to the right of 2002. This behavior suggests a quadratic function.
   b. $p(t) \approx 0.038t^2 - 0.079t + 3.532$ billion passengers gives the number of enplaned passengers worldwide t years since 2000, data from $0 \le t \le 7$.
   c. no; The number of enplaned passengers should continue to increase, but it makes more sense to assume it will follow the linear trend suggested after 2003.

Chapter 2 Describing Change: Rates

Section 2.1 Measuring Change over an Interval (pages 134–138)

1. $\text{AROC} = \dfrac{\text{change}}{\text{length of interval}}$

   $= \dfrac{151.80 - 156.86}{50}$

   $= -0.1012$

   During a media event at which Steve Jobs spoke, Apple shares dropped by an average of $0.10 per minute.
   The result could also be reported as 10¢ per minute.

3. $\text{AROC} = \dfrac{\text{change}}{\text{length of interval}}$

   $= \dfrac{3660 - 1856}{9}$

   ≈ 200.444

   The average community college tuition in Iowa increased by an average of $200.44 per year from 2000-2001 to 2009-2010.

5. a. change = new − old

   $= 17.6 - (-121.6)$

   $= \$139.2 \text{ million}$

   Between the end of 2008 and the end of 2009, AirTran's profit increased by $139.2 million.

   b. Percentage change does not apply because the input interval spans both positive and negative numbers and therefore does not have an absolute scale against which to measure percentage change.

   c. $\text{AROC} = \dfrac{\text{change}}{\text{length of interval}}$

   $= \dfrac{139.2}{2009 - 2008}$

   $= \$139.2 \text{ per year}$

   Between the end of 2008 and the end of 2009, AirTran's profit increased by $139.2 million per year.

7. a. change = new − old

   $= 43 - 40$

   $= 3 \text{ percentage points}$

Between 2004 and 2008, the percentage of students meeting national mathematics benchmarks increased by 3 percentage points.

b. percentage change $= \dfrac{\text{change}}{\text{old}} \cdot 100\%$

$= \dfrac{3}{40} \cdot 100\%$

$\approx \boxed{7.5\%}$

Between 2004 and 2008, the percentage of students meeting national mathematics benchmarks increased by 7.5%.

c. $\text{AROC} = \dfrac{\text{change}}{\text{length of interval}}$

$= \dfrac{3}{2004 - 2000}$

$= \boxed{0.75 \text{ percentage points per year}}$

Between 2004 and 2008, the percentage of students meeting national mathematics benchmarks increased by 0.75 percentage points per year.

9. (The figure in the activity erroneously identifies the 13th trading day as being October 22. It should be October 19.)

a. percentage change $= \dfrac{\text{new} - \text{old}}{\text{old}} \cdot 100\%$

$= \dfrac{303.4 - 193.2}{193.2} \cdot 100\%$

$\approx \boxed{57.039\%}$

$\text{AROC} = \dfrac{\text{change}}{\text{length of interval}}$

$= \dfrac{303.4 - 193.2}{22 - 1}$

$\approx \boxed{5.248 \text{ million shares per day}}$

b.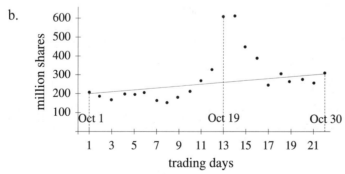

c. (The answer key erroneously identifies the 13th trading day as October 13.)
Answers will vary but should include a statement indicating that the spike near October 19 is not described adequately by the slope of the secant line.

11. a. $\text{AROC} = \dfrac{\text{change}}{\text{length of interval}}$

$= \dfrac{5682 - 4860}{7 - 4}$

$\approx \boxed{274 \text{ million dollars per year}}$

$\boxed{\text{Between 2004 and 2007, Kelly's sales of service increased by an average of \$274 million per year.}}$

b. $\text{percentage change} = \dfrac{\text{new} - \text{old}}{\text{old}} \cdot 100\%$

$= \dfrac{5514 - 5682}{5682} \cdot 100\%$

$\approx \boxed{-2.957\%}$

$\boxed{\text{Between 2007 and 2008, Kelly's sales of service decreased by 2.957\%.}}$

c. $\text{change} = \text{new} - \text{old}$

$= 5514 - 4860$

$= \boxed{\$654 \text{ million}}$

13. a. $P(t) \approx -0.037t^2 + 25.529t - 527.143$ thousand dollars profit where t dollars is the ticket price, data from $200 \leq t \leq 450$.

b. $\dfrac{P(350) - P(200)}{350 - 200} \approx \boxed{4.943 \text{ thousand dollars per dollar}}$

(thousand dollars of profit per dollar of ticket price)

c. $\dfrac{P(450) - P(350)}{450 - 350} \approx \boxed{-4.414 \text{ thousand dollars per dollar}}$

15. a. $\dfrac{12.3 - 69.7}{70 - 0} = \boxed{-0.82 \text{ years per year}}$ (years of life expectancy per year of age)

b. between the ages of 10 and 20: $\dfrac{51.3 - 60.9}{20 - 10} = -0.96$ years per year

between the ages of 20 and 30: $\dfrac{42.4 - 51.3}{30 - 20} = -0.89$ years per year

$\boxed{\text{Life expectancy decreases more rapidly between ages 10 and 20 than between ages 20 and 30.}}$

17. a. $\dfrac{u(4) - u(0)}{4 - 0} \approx \boxed{1.752 \text{ million users per year}}$

b. $\dfrac{u(4) - u(0)}{u(0)} \cdot 100\% \approx \boxed{87.389\%}$

c. $\dfrac{u(4)}{109.9554} \approx \boxed{13.668\%}$

19. a. $\dfrac{s(13)-s(3)}{13-3} \approx$ $\boxed{\$0.107 \text{ per year}}$

$\boxed{\text{Between 1998 and 2008, the ATM surcharge for non-account holders increased by an average of \$0.11 per year.}}$ (This result can also be expressed as 11¢ per year.)

b. change $= s(13) - s(3) = \boxed{\$1.07}$

percentage change $= \dfrac{s(13)-s(3)}{s(3)} \cdot 100\% \approx \boxed{117.900\%}$

21. a. i. AROC $= \dfrac{f(3)-f(1)}{3-1} = \boxed{3}$

percentage change $= \dfrac{f(3)-f(1)}{f(1)} \cdot 100\% \approx \boxed{85.714\%}$

ii. AROC $= \dfrac{f(5)-f(3)}{5-3} = \boxed{3}$

percentage change $= \dfrac{f(5)-f(3)}{f(3)} \cdot 100\% \approx \boxed{46.154\%}$

iii. AROC $= \dfrac{f(7)-f(5)}{7-5} = \boxed{3}$

percentage change $= \dfrac{f(7)-f(5)}{f(5)} \cdot 100\% \approx \boxed{31.579\%}$

b. For a linear function, the average rate of change between two points is constant, but the percentage change varies.

23. a. Answers will vary but might be include the following:
It is not possible to find the mid-year balance without fitting an exponential model to the data.

b. $f(t) \approx 1400(1.064^t)$ dollars gives the balance after t years on an account with an initial deposit of \$1400 and continuous compounding.

percentage change $= \dfrac{f(4)-f(3.5)}{f(3.5)} \cdot 100\% \approx \boxed{3.149\%}$

AROC $= \dfrac{f(4)-f(3.5)}{4-3.5} \approx \boxed{109.524 \text{ dollars per year}}$

Section 2.2 Measures of Change at a Point—Graphical (pages 147–152)

1. a. $\boxed{\text{negative at } A}$ because the function is decreasing at that point
$\boxed{\text{zero at } B \text{ and } D}$ because the function is continuous and changes direction at those points
$\boxed{\text{positive at } C \text{ and } E}$ because the function is increasing at those point

b. The function is steeper at $\boxed{E}$ because the tangent line at E is closer to being vertical than is the tangent line at C.

c. The function is steeper at $\boxed{A}$ because the tangent line at A is closer to being vertical than is the tangent line at C.

3. a. increasing over ⬚none⬚ of the input interval shown
    b. decreasing over the ⬚entire input interval⬚ shown
    c. $x \approx -1$

5. a. ⬚negative slope over all x shown⬚ because the function is decreasing
    b. ⬚constant slope over all x shown⬚ because the function is linear

7. a. ⬚negative slope over all x shown⬚ because the function is decreasing
    b. ⬚steepness decreases⬚ because lines tangent to the graph become closer to being horizontal

9. a. At 60% humidity, the pupae graph is decreasing so has negative slope. Both the ⬚eggs⬚ and ⬚larvae⬚ graphs are increasing, that is, they have positive slopes.
    b. The graphs for ⬚eggs and larvae⬚ are both increasing at 60% relative humidity, showing an increase in survival rate. The graph for ⬚pupae⬚ is decreasing at 60% relative humidity, showing a decrease in survival rate.
    c. At 97% relative humidity, the graph for eggs is steeper than the graph for pupae. Since the graphs are both decreasing, the slope of the line tangent to the ⬚eggs⬚ graph is more negative than the slope of the line tangent to the ⬚pupae⬚ graph.
    d. Because the graph for ⬚pupae⬚ is the only graph of the three that is decreasing over the entire input interval for which it is shown, it is the only graph that satisfies the "any tangent lines" condition.
    e. Because the graph for ⬚larvae⬚ is the only graph of the three that is increasing over the entire input interval for which it is shown, it is the only graph that satisfies the "always positive" condition.
    f. At 30% relative humidity, the graphs for ⬚eggs⬚ and ⬚pupae⬚ are both decreasing and appear to have similar steepness.
    g. At 65% relative humidity, the graphs for ⬚eggs⬚ and ⬚larvae⬚ are both increasing and appear to have similar steepness.

11. Answers will vary depending on the estimation of the point in part *a*.
    a. (4, 9250)
    b. $\dfrac{9400 - 9250}{4.8 - 4} = $ ⬚187.5 thousand employees per year⬚
    c. $\dfrac{187.5}{9400} \cdot 100\% \approx$ ⬚1.995% per year⬚

13. Answers will vary depending on the estimation of the point in part *a*.
    a. (0.5, 60)
    b. $\dfrac{60 - 40}{0.5 - 5.5} =$ ⬚−4 thousand feet per inch⬚
    c. $\dfrac{-4}{40} \cdot 100\% =$ ⬚−10% per inch⬚

15. The lines at ⬚A and C are not tangent to the graph⬚ because they do not have the same steepness as the graph.

17.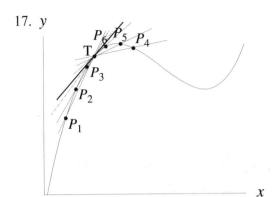

19. a. At point A, the function is concave up and the tangent line lies below the graph.
    At point B, the function has a change of concavity and the tangent line cuts through the graph.
    At point C, the function is concave down and the tangent line lies above the graph.
    At point D, the function is linear and the tangent line coincides with the graph.
    b. (The answer key erroneously identifies C as having positive slope.)
    At points A, and B the slope is positive because the function is increasing at those points.
    At point C, the slope is zero.
    At point D, the slope is negative because the function is decreasing at that point.
    c.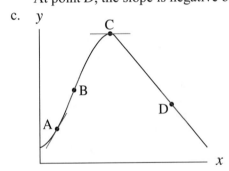

21. a.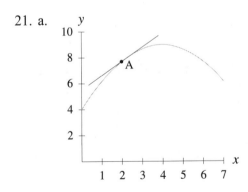

    b. Using points (2, 7.7) and (3, 9.0) estimated from the graph,
    $$\text{slope} = \frac{9.0 - 7.7}{3 - 2} = \boxed{1.3}.$$

23. a.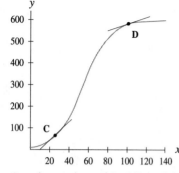

b. C: using points (20, 40) and (40, 120) estimated from the graph,
$$\text{slope} = \frac{120-40}{40-20} = \boxed{4}.$$
D: using points (100, 580) and (120, 610) estimated from the graph,
$$\text{slope} = \frac{610-580}{120-100} = \boxed{1.5}.$$

25. a.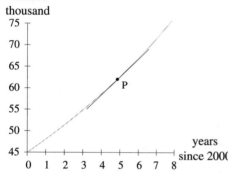

Using points (4, 59) and (5, 63) estimated from the graph, the estimate of the slope is 4.
$$\text{slope} = \frac{63-59}{5-4} = \boxed{4}.$$

b. hundred thousand subscribers per year; In 2005, the total number of cellular subscribers was increasing by 4 hundred thousand per year.

c. $\frac{4}{63} \cdot 100\% \approx \boxed{6.349\% \text{ per year}}$; In 2005, the total number of cellular subscribers was increasing by approximately 6.3% per year.

27. a.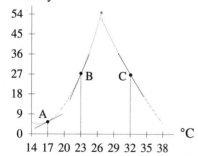

A: using points $(17, 5.5)$ and $(19, 8.5)$ estimated from the graph,
$$\text{slope} = \frac{8.5 - 5.5}{19 - 17} = \boxed{1.5 \text{ mm/day per } °C}.$$
B: using points $(23, 27)$ and $(24, 32.5)$ estimated from the graph,
$$\text{slope} = \frac{32.5 - 27}{24 - 23} = \boxed{5.5 \text{ mm/day per } °C}.$$
C: using points $(32, 27)$ and $(34, 18)$ estimated from the graph,
$$\text{slope} = \frac{18 - 27}{34 - 32} = \boxed{-4.5 \text{ mm/day per } °C}.$$

b. At a temperature of 32 °C, the growth of a pea seedling is decreasing by 4.5 mm/day per °C.

c. $\frac{-4.5}{27} \cdot 100\% \approx \boxed{16.667\% \text{ per } °C}$; $\boxed{\text{At a temperature of 32 °C, the growth of a pea seedling is decreasing by 16.667\% per °C.}}$

29. Answers will vary but might include the following ideas:
Average rate of change is graphically represented as the slope of the secant line between two points. The slope of a tangent line at a point is used to find the rate of change at the point.

Section 2.3 Rates of Change—Notation and Interpretation (pages 157–159)

1. a. units of measure for output of $p' = \frac{\text{output units of } p}{\text{input unit of } p}$

    $= \frac{\text{miles}}{\text{hour}}$

    $= \boxed{\text{miles per hour}}$

   b. The rate of change of distance with respect to time is known as the $\boxed{\text{speed}}$.

3. a. When the ticket price is $65, the weekly profit to the airline on flights from Boston to Washington is $15,000.
   b. When the ticket price is $65, the weekly profit to the airline on flights from Boston to Washington is increasing by $1.5 thousand per dollar (of ticket price).
   c. When the ticket price is $90, the weekly profit to the airline on flights from Boston to Washington is decreasing by $2 thousand per dollar (of ticket price).

5. a. $w(2)$ cannot be negative because the lowest number of words typed in a minute is 0.

   b. units of measure for output of $w' = \frac{\text{output units of } w}{\text{input unit of } w}$

    $= \frac{\text{words per minute}}{\text{week}}$

    $= \boxed{\text{wpm per week}}$

   c. $\left.\frac{dw}{dt}\right|_{t=2}$ can be negative if the number of words typed per minute is decreasing at week 2.

7. a. $P(30)$ can be negative if the cost of the 30 shirts is greater than the revenue from the sales of 30 shirts.
   b. $P'(100)$ can be negative if the profit is decreasing as additional shirts beyond 100 shirts are sold.
   c. If $P'(200)$ is negative, the fraternity's profit is decreasing. More information is necessary to determine whether the fraternity is "losing money".

9. One possible graph:

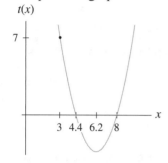

11. a. At the beginning of a diet, a person weighed 167 pounds.
       After 12 weeks on the diet, that person weighed 142.
    b. After one week on the diet, the person's weight was decreasing by 2 pounds per week.
       After 9 weeks on the diet, the person's weight was decreasing by 1 pound per week.
    c. After 12 weeks on the diet, the person's weight was not changing.
       After 15 weeks on the diet, the person's weight is increasing by 0.25 pounds per week.
    d. One possible graph:

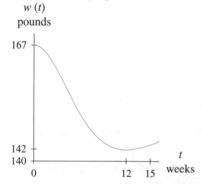

13. a. units of measure for output of $D' = \dfrac{\text{output units of } D}{\text{input unit of } D}$

    $= \dfrac{\text{years}}{\text{percentage point}}$

    $= \boxed{\text{years per percentage point}}$

    b. $\left.\dfrac{dD}{dr}\right|_{r=a}$ is negative for every positive value of r because for larger interest rates the investment will double more quickly. Doubling time decreases as r increases so the rate of change is negative.
    c. i. At 9% interest compounded continuously, an investment will double its value in 7.7 years.

ii. If the interest rate for an investment at 5% compounded continuously is changed to 6% compounded continuously, the doubling time will decrease by approximately 2.77 years.
iii. If the interest rate for an investment of 12% compounded continuously is changed to 13%, the doubling time will decrease by approximately 0.48 year.
iv. At 16% interest compounded continuously, an investment will double its value in 5.79 years.

15. One possible graph:

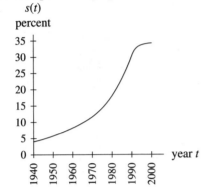

17. a.
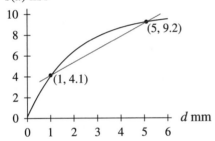

Using the points (1, 4.1) and (5, 9.2) estimated from the graph,

$$\text{slope} = \frac{9.2 - 4.1}{5 - 1} = \boxed{1.275 \text{ m/s per mm}}.$$

The slope gives the average rate of change in the terminal speed of a raindrop of diameter d as diameter changes from 1mm to 5mm.

b. The slope of the tangent line at $d = 4$ is an estimate of the rate of change of the terminal speed of a raindrop of size 4 mm as the diameter of the raindrop increases.

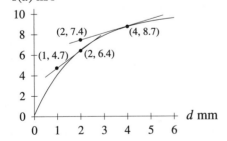

c. Using points (4, 8.7) and (2, 7.4) estimated from the graph,

$$\text{slope} = \frac{8.7 - 7.4}{4 - 2} = \boxed{0.65 \text{ m/s per mm}}.$$

The terminal speed of a falling raindrop with diameter 4 mm is increasing by 0.65 m/s per mm.

d. Using the points (2, 6.4) and (1, 4.7) estimated from the graph,

$$\text{slope} = \frac{6.4 - 4.7}{2 - 1} = 1.7$$

$$\%\text{ROC} = \frac{1.7}{6.4} \cdot 100\% \approx \boxed{26.563\% \text{ per mm}}$$

The terminal speed of a falling raindrop with diameter 2 mm is increasing by 26.563% per mm.

19. a.

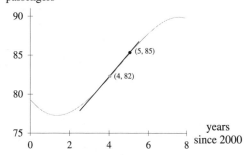

Using the points (4, 82) and (5, 85) estimated from the line tangent to the graph at $x = 4$,

$$\text{slope} = \frac{85 - 82}{5 - 4} = \boxed{3 \text{ million passengers per year}}$$

b. $\%\text{ROC} = \frac{3}{82} \cdot 100\% \approx \boxed{3.659\% \text{ per year}}$

c. In 2004, the number of passengers going through Hartsfield-Jackson Atlanta International Airport was increasing by 3 million passengers per year or 3.659% per year.

Section 2.4 Rates of Change—Numerical Limits and Nonexistence (pages 163–167)

1.

$x \to 2^-$	$f(x)$	Slope of secant $= \dfrac{f(2)-f(x)}{2-x}$	$x \to 2^+$	$f(x)$	Slope of secant $= \dfrac{f(2)-f(x)}{2-x}$
1.9	3.732132	$\dfrac{4-3.732132}{2-1.9} \approx 2.679$	2.1	4.287094	$\dfrac{4-4.287094}{2-2.1} \approx 2.871$
1.99	3.972370	$\dfrac{4-3.972370}{2-1.99} \approx 2.763$	2.01	4.027822	$\dfrac{4-4.027822}{2-2.01} \approx 2.782$
1.999	3.997228	$\dfrac{4-3.997228}{2-1.999} \approx 2.772$	2.001	4.002773	$\dfrac{4-4.002773}{2-2.001} \approx 2.774$
1.9999	3.999723	$\dfrac{4-3.999723}{2-1.9999} \approx 2.773$	2.0001	4.000277	$\dfrac{4-4.000277}{2-2.0001} \approx 2.773$
1.99999	3.999972	$\dfrac{4-3.999972}{2-1.99999} \approx 2.773$	2.00001	4.000028	$\dfrac{4-4.000028}{2-2.00001} \approx 2.773$
1.999999	3.999997	$\dfrac{4-3.999997}{2-1.999999} \approx 2.773$	2.000001	4.000003	$\dfrac{4-4.000003}{2-2.000001} \approx 2.773$
		$\lim\limits_{x \to 2^-}\left(\text{slopes of secants}\right) \approx 2.8$			$\lim\limits_{x \to 2^+}\left(\text{slopes of secants}\right) \approx 2.8$

$f'(2) = \lim\limits_{x \to 2}\left(\text{slopes of secants}\right) = \lim\limits_{x \to 2} \dfrac{f(2)-f(x)}{2-x} \approx \boxed{2.8}$

(The result is rounded to the precision requested in the activity. Even though 2.773 is correct to three decimals, the numerical estimation chart does not conclusively support such precision.)

3.

$x \to 1^-$	$f(x)$	Slope of secant $= \dfrac{f(1)-f(x)}{1-x}$	$x \to 1^+$	$f(x)$	Slope of secant $= \dfrac{f(1)-f(x)}{1-x}$
0.9	1.897367	$\dfrac{2-1.897367}{1-0.9} \approx 1.0263$	1.1	2.097618	$\dfrac{2-2.097618}{1-1.1} \approx 0.9762$
0.99	1.989975	$\dfrac{2-1.989975}{1-0.99} \approx 1.0025$	1.01	2.009975	$\dfrac{2-2.009975}{1-1.01} \approx 0.9975$
0.999	1.999000	$\dfrac{2-1.999000}{1-0.999} \approx 1.0003$	1.001	2.001000	$\dfrac{2-2.001000}{1-1.001} \approx 0.9998$
0.9999	1.999900	$\dfrac{2-1.999900}{1-0.9999} \approx 1.0000$	1.0001	2.000100	$\dfrac{2-2.000100}{1-1.0001} \approx 1.0000$
0.99999	1.999990	$\dfrac{2-1.999990}{1-0.99999} \approx 1.0000$	1.00001	2.000010	$\dfrac{2-2.000010}{1-1.00001} \approx 1.0000$
0.999999	1.999999	$\dfrac{2-1.999999}{1-0.999999} \approx 1.0000$	1.000001	2.000001	$\dfrac{2-2.000001}{1-1.000001} \approx 1.0000$
		$\lim\limits_{x \to 1^-}\left(\text{slopes of secants}\right) \approx 1.00$			$\lim\limits_{x \to 1^+}\left(\text{slopes of secants}\right) \approx 1.00$

$f'(1) = \lim_{x \to 1}(\text{slopes of secants}) = \lim_{x \to 1}\frac{f(1)-f(x)}{1-x} \approx \boxed{1.00}$

5. a.

$t \to 6^-$	Slope of secant $= \frac{p(6)-p(t)}{6-t}$	$t \to 6^+$	Slope of secant $= \frac{p(6)-p(t)}{6-t}$
5.9	$\frac{p(6)-p(5.9)}{6-5.9} = 2.41758$	6.1	$\frac{p(6)-p(6.1)}{6-6.1} = 2.32638$
5.99	$\frac{p(6)-p(5.99)}{6-5.99} = 2.37845$	6.01	$\frac{p(6)-p(6.01)}{6-6.01} = 2.36953$
5.999	$\frac{p(6)-p(5.999)}{6-5.999} \approx 2.37445$	6.001	$\frac{p(6)-p(6.001)}{6-6.001} \approx 2.37355$
5.9999	$\frac{p(6)-p(5.9999)}{6-5.9999} \approx 2.37404$	6.0001	$\frac{p(6)-p(6.0001)}{6-6.0001} \approx 2.37396$
5.99999	$\frac{p(6)-p(5.99999)}{6-5.99999} \approx 2.37400$	6.00001	$\frac{p(6)-p(6.00001)}{6-6.00001} \approx 2.37400$
5.999999	$\frac{p(6)-p(5.999999)}{6-5.999999} \approx 2.37400$	6.000001	$\frac{p(6)-p(6.000001)}{6-6.000001} \approx 2.37400$
5.9999999	$\frac{p(6)-p(5.9999999)}{6-5.9999999} \approx 2.37400$	6.0000001	$\frac{p(6)-p(6.0000001)}{6-6.0000001} \approx 2.37400$
	$\lim_{t \to 6^-}\left(\text{slopes of secants}\right) \approx 2.374$		$\lim_{t \to 6^+}\left(\text{slopes of secants}\right) \approx 2.374$

$p'(6) = \lim_{t \to 6}(\text{slopes of secants}) = \lim_{t \to 6}\frac{p(6)-p(t)}{6-t} \approx \boxed{2.374}$

At the end of 2006, the number of passengers going through Hartsfield-Jackson International Airport was increasing by approximately 2,374 thousand per year.

b. Using the rounded numerical estimate for $p'(6)$,

$\%\text{ROC} = \frac{p'(6)}{p(6)} \cdot 100\% \approx \boxed{2.713\% \text{ per year}}$

At the end of 2006, the number of passengers going through Hartsfield-Jackson International Airport was increasing by approximately 2.713% per year.

2.4 Rates of Change—Numerical Limits and Nonexistence
Solutions to Odd-Numbered Activities

7. a.

$x \to 13^-$	Slope of secant $= \dfrac{t(13)-t(x)}{13-x}$	$x \to 13^+$	Slope of secant $= \dfrac{t(13)-t(x)}{13-x}$
12.9	$\dfrac{t(13)-t(12.9)}{13-12.9} \approx -3.775$	13.1	$\dfrac{F(13)-F(13.1)}{13-13.1} \approx -3.739$
12.99	$\dfrac{t(13)-t(12.99)}{13-12.99} \approx -3.759$	13.01	$\dfrac{F(13)-F(13.01)}{13-13.01} \approx -3.755$
12.999	$\dfrac{t(13)-t(12.999)}{13-12.999} \approx -3.757$	13.001	$\dfrac{F(13)-F(13.001)}{13-13.001} \approx -3.757$
12.9999	$\dfrac{t(13)-t(12.9999)}{13-12.9999} \approx -3.757$	13.0001	$\dfrac{F(13)-F(13.0001)}{13-13.0001} \approx -3.757$
12.99999	$\dfrac{t(13)-t(12.99999)}{13-12.99999} \approx -3.757$	13.00001	$\dfrac{F(13)-F(13.00001)}{13-13.00001} \approx -3.757$
	$\lim\limits_{x \to 13^-}\begin{pmatrix}\text{slopes of}\\\text{secants}\end{pmatrix} \approx -3.8$		$\lim\limits_{x \to 13^+}\begin{pmatrix}\text{slopes of}\\\text{secants}\end{pmatrix} \approx -3.8$

$$t'(13) = \lim_{x \to 13}(\text{slopes of secants}) = \lim_{x \to 13} \frac{t(13)-t(x)}{13-x} \approx \boxed{-3.8 \text{ seconds per year (of age)}}$$

b. Using the rounded numerical estimate for $t'(13)$,

$$\%\text{ROC} = \frac{t'(13)}{t(13)} \cdot 100\% \approx \boxed{-5.6\% \text{ per year (of age)}}.$$

c. The negative rate of change shows that swim time is decreasing. The swimmer's time is $\boxed{\text{improving}}$.

9. a.

$t \to 8^-$	Slope of secant $= \dfrac{S(8)-S(t)}{8-t}$	$t \to 8^+$	Slope of secant $= \dfrac{S(8)-S(t)}{8-t}$
7.9	$\dfrac{S(8)-S(7.9)}{8-7.9} \approx 4.727$	8.1	$\dfrac{S(8)-S(8.1)}{8-8.1} \approx 4.010$
7.99	$\dfrac{S(8)-S(7.99)}{8-7.99} \approx 4.408$	8.01	$\dfrac{S(8)-S(8.01)}{8-8.01} \approx 4.336$
7.999	$\dfrac{S(8)-S(7.999)}{8-7.999} \approx 4.376$	8.001	$\dfrac{S(8)-S(8.001)}{8-8.001} \approx 4.368$
7.9999	$\dfrac{S(8)-S(7.9999)}{8-7.9999} \approx 4.372$	8.0001	$\dfrac{S(8)-S(8.0001)}{8-8.0001} \approx 4.372$
7.99999	$\dfrac{S(8)-S(7.99999)}{8-7.99999} \approx 4.372$	8.00001	$\dfrac{S(8)-S(8.00001)}{8-8.00001} \approx 4.372$
7.999999	$\dfrac{S(8)-S(7.999999)}{8-7.999999} \approx 4.372$	8.000001	$\dfrac{S(8)-S(8.000001)}{8-8.000001} \approx 4.372$
	$\lim\limits_{t \to 8^-}\begin{pmatrix}\text{slopes of}\\\text{secants}\end{pmatrix} \approx 4.4$		$\lim\limits_{t \to 8^+}\begin{pmatrix}\text{slopes of}\\\text{secants}\end{pmatrix} \approx 4.4$

$$S'(8) = \lim_{t \to 8}(\text{slopes of secants}) = \lim_{t \to 8} \frac{S(8)-S(t)}{8-t} \approx \boxed{4.4 \text{ billion dollars per year}}$$

b. At the end of 2008, annual U.S. factory sales of consumer electronics goods to dealers were increasing by approximately $4.4 billion per year.

11. a. A scatter plot suggests a single concavity with a change in direction (relative maximum).
b. $s(t) \approx -2.814t^2 + 41.469t - 82.586$ thousand dollars gives the average weekly sales for Abercrombie and Fitch where x is the number of years since 2000, data from 2004 through 2008.
c.

$t \to 7^-$	Slope of secant $= \dfrac{s(7)-s(t)}{7-t}$	$t \to 7^+$	Slope of secant $= \dfrac{s(7)-s(t)}{7-t}$
6.9	$\dfrac{s(7)-s(6.9)}{7-6.9} \approx 2.351$	7.1	$\dfrac{s(7)-s(7.1)}{7-7.1} \approx 1.788$
6.99	$\dfrac{s(7)-s(6.99)}{7-6.99} \approx 2.098$	7.01	$\dfrac{s(7)-s(7.01)}{7-7.01} \approx 2.041$
6.999	$\dfrac{s(7)-s(6.999)}{7-6.999} \approx 2.072$	7.001	$\dfrac{s(7)-s(7.001)}{7-7.001} \approx 2.066$
6.9999	$\dfrac{s(7)-s(6.9999)}{7-6.9999} \approx 2.070$	7.0001	$\dfrac{s(7)-s(7.0001)}{7-7.0001} \approx 2.069$
6.99999	$\dfrac{s(7)-s(6.99999)}{7-6.99999} \approx 2.070$	7.00001	$\dfrac{s(7)-s(7.00001)}{7-7.00001} \approx 2.069$
6.999999	$\dfrac{s(7)-s(6.999999)}{7-6.999999} \approx 2.069$	7.000001	$\dfrac{s(7)-s(7.000001)}{7-7.000001} \approx 2.069$
	$\lim_{t \to 7^-}\left(\text{slopes of secants}\right) \approx 2.1$		$\lim_{t \to 7^+}\left(\text{slopes of secants}\right) \approx 2.1$

$s'(7) = \lim_{t \to 7}\left(\text{slopes of secants}\right) = \lim_{t \to 7}\dfrac{s(7)-s(t)}{7-t} \approx$ $2.1 thousand per year

d. In 2007, the average weekly sales for Abercrombie and Fitch were increasing by $2,100 per year.

13. a. $r(x) \approx \dfrac{1.937}{1+29.064e^{-0.421x}}$ U/100μL gives a measurement of the reaction activity of a chemical mixture x minutes after the mixture reaches a temperature of $95°C$, data from $0 \le x \le 18$. The limiting value of the logistic function is 1.937 U/100μL.

b. AROC $= \dfrac{r(11)-r(7)}{11-7} \approx$ 0.186 U/100μL per minute

c.

$x \to 9^-$	Slope of secant $= \dfrac{r(9)-r(x)}{9-x}$	$x \to 9^+$	Slope of secant $= \dfrac{r(9)-r(x)}{9-x}$
8.9	$\dfrac{r(9)-r(8.9)}{9-8.9} \approx 0.19598$	9.1	$\dfrac{r(9)-r(9.1)}{9-9.1} \approx 0.19428$
8.99	$\dfrac{r(9)-r(8.99)}{9-8.99} \approx 0.19524$	9.01	$\dfrac{r(9)-r(9.01)}{9-9.01} \approx 0.19507$
8.999	$\dfrac{r(9)-r(8.999)}{9-8.999} \approx 0.19517$	9.001	$\dfrac{r(9)-r(9.001)}{9-9.001} \approx 0.19515$
8.9999	$\dfrac{r(9)-r(8.9999)}{9-8.9999} \approx 0.19516$	9.0001	$\dfrac{r(9)-r(9.0001)}{9-9.0001} \approx 0.19516$
8.99999	$\dfrac{r(9)-r(8.99999)}{9-8.99999} \approx 0.19516$	9.00001	$\dfrac{r(9)-r(9.00001)}{9-9.00001} \approx 0.19516$
8.999999	$\dfrac{r(9)-r(8.999999)}{9-8.999999} \approx 0.19516$	9.000001	$\dfrac{r(9)-r(9.000001)}{9-9.000001} \approx 0.19516$
	$\lim\limits_{x \to 9^-}(\text{slopes of secants}) \approx 0.195$		$\lim\limits_{x \to 9^+}(\text{slopes of secants}) \approx 0.195$

$$r'(9) = \lim_{x \to 9}(\text{slopes of secants}) = \lim_{x \to 9} \frac{r(9)-r(x)}{9-x} \approx \boxed{0.195 \text{ U}/100\mu\text{L per minute}}$$

d. Between 7 and 11 minutes after the mixture reached a temperature of $95°C$, the reaction activity increased by an average of 0.186 U/100μL per minute.
After 9 minutes the reaction activity was increasing by 0.195 U/100μL per minute.

15. The function is continuous on the input interval shown, but it is not differentiable at $x=4$ because the limits of the slopes of secant lines from the right and left do not equal each other at $x=4$.

17. The function is not continuous (and therefore not differentiable) at $x=1$. There is a vertical asymptote at $x=1$.

19. a. The line tangent to the graph is not defined for $x = \boxed{8}$.
    b. The graph appears to be $\boxed{\text{continuous}}$ at $x=8$.
       $\boxed{\text{This graph could be traced without lifting the pencil from the page.}}$
    c. The graph appears to have a sharp corner at $x=8$. The limit of secants from the right does not equal the limit of secants from the left at this point.

21. a. The line tangent to the graph is not defined for $t = \boxed{26}$.
    b. The graph is $\boxed{\text{not continuous}}$ for $t=26$.
       $\boxed{\text{This graph cannot be traced without lifting the pencil from the page at that input value.}}$
    c. The tangent line is not defined for $t=26$ because the graph is not continuous at that point.

23. Answers will vary but might be similar to the following:
    Rates of change can be quickly estimated graphically when a graph with a grid or numbers on the axes is provided. Rates of change can be estimated numerically when an equation is provided. Numerical estimation is generally much more accurate but also more time consuming than graphical estimation.

25. Answers will vary but might be similar to the following:
Continuous piecewise functions have a tangent line at their break point if the limit of the slope of the secants from the left is equal to the limit of the slope of the secants from the right.

$$\lim_{x \to 2^-}\left(\text{slope of the secants of } f(x)\right) = -4 \text{ and } \lim_{x \to 2^+}\left(\text{slope of the secants of } f(x)\right) = 3 \text{ so a tangent could not be drawn at } x = 2.$$

$$\lim_{x \to 3^-}\left(\text{slope of the secants of } g(x)\right) = 27 \text{ and } \lim_{x \to 3^+}\left(\text{slope of the secants of } f(x)\right) = 27 \text{ so a tangent could be drawn at } x = 3$$

Section 2.5 Rates of Change Defined over Intervals (pages 172–174)

1. typical point: $(x,\ 3x-2)$
   close point: $(x+h,\ 3(x+h)-2) = (x+h,\ 3x+3h-2)$

   $$\frac{df}{dx} = \lim_{h \to 0}\frac{(3x+3h-2)-(3x-2)}{h}$$

   $$= \lim_{h \to 0}\frac{3h}{h}$$

   $$= \lim_{h \to 0} 3 \qquad \text{by the Replacement Rule}$$

   $$= 3$$

3. typical point: $(x,\ 3x^2)$
   close point: $(x+h,\ 3(x+h)^2) = (x+h,\ 3x^2 + 6xh + 3h^2)$

   $$f'(x) = \lim_{h \to 0}\frac{(3x^2 + 6xh + 3h^2) - 3x^2}{h}$$

   (continued on the next page)

   $$= \lim_{h \to 0}\frac{6xh + 3h^2}{h} \qquad \left(\text{or } \lim_{h \to 0}\frac{h(6x+3h)}{h}\right)$$

   $$= \lim_{h \to 0}(6x + 3h) \qquad \text{by the Replacement Rule}$$

   $$= 6x$$

5. typical point: $(x,\ x^3)$
   close point: $(x+h,\ (x+h)^3) = (x+h,\ x^3 + 3x^2h + 3xh^2 + h^3)$

   $$f'(x) = \lim_{h \to 0}\frac{(x^3 + 3x^2h + 3xh^2 + h^3) - x^3}{h}$$

   $$= \lim_{h \to 0}\frac{3x^2h + 3xh^2 + h^3}{h} \qquad \left(\text{or } \lim_{h \to 0}\frac{h(3x^2 + 3xh + h^2)}{h}\right)$$

   $$= \lim_{h \to 0}(3x^2 + 3xh + h^2) \qquad \text{by the Replacement Rule}$$

   $$= 3x^2$$

2.6 Rate-of-Change Graph
Solutions to Odd-Numbered Activities

7. a. typical point: $(x, 4x^2)$

   close point: $(x+h, 4(x+h)^2) = (x+h, 4x^2 + 8xh + 4h^2)$

   $$f'(x) = \lim_{h \to 0} \frac{(\cancel{4x^2} + 8xh + 4h^2) - \cancel{4x^2}}{h}$$

   $$= \lim_{h \to 0} \frac{8xh + 4h^2}{h} \quad \left(\text{or} \quad \lim_{h \to 0} \frac{h(8x + 4h)}{h}\right)$$

   $$= \lim_{h \to 0}(8x + 4h) \quad \text{by the Replacement Rule}$$

   $$= 8x$$

   $\boxed{f'(x) = 8x}$

   b. $f'(2) = \boxed{16}$

9. a. typical point: $(t, 4t^2 - 3)$

   close point: $(t+h, 4(t+h)^2 - 3) = (t+h, 4t^2 + 8th + 4h^2 - 3)$

   $$g'(t) = \lim_{h \to 0} \frac{(\cancel{4t^2} + 8th + 4h^2 - \cancel{3}) - (\cancel{4t^2} - \cancel{3})}{h}$$

   $$= \lim_{h \to 0} \frac{8th + 4h^2}{h} \quad \left(\text{or} \quad \lim_{h \to 0} \frac{h(8t + 4h)}{h}\right)$$

   $$= \lim_{h \to 0}(8t + 4h) \quad \text{by the Replacement Rule}$$

   $$= 8t$$

   $\boxed{g'(t) = 8t}$

   b. $g'(4) = \boxed{32}$

11. (Instead of using h as the function name as in the text use f as the function name to avoid confusion. The function is $f(t) = -16t^2 + 100$.)

    a. typical point: $(t, -16t^2 + 100)$

    close point: $(t+h, -16(t+h)^2 + 100) = (t+h, -16t^2 - 32th - 16h^2 + 100)$

    $$f'(t) = \lim_{h \to 0} \frac{(\cancel{-16t^2} - 32th - 16h^2 + \cancel{100}) - (\cancel{-16t^2} + \cancel{100})}{h}$$

    $$= \lim_{h \to 0} \frac{-32th - 16h^2}{h} \quad \left(\text{or} \quad \lim_{h \to 0} \frac{h(-32t - 16h)}{h}\right)$$

    $$= \lim_{h \to 0}(-32t - 16h) \quad \text{by the Replacement Rule}$$

    $$= -32t$$

    $\boxed{f'(t) = -32t \text{ feet per second}}$

    b. $f'(1) = \boxed{-32}$ feet per second

    $\boxed{\text{After 1 second, the object is falling at a rate of 32 feet per second.}}$

13. a. typical point: $\left(t,\ 1.2t^2 - 6.1t + 39.5\right)$

   close point: $\left(t+h,\ 1.2(t+h)^2 - 6.1(t+h) + 39.5\right)$
   $= \left(t+h,\ 1.2t^2 + 2.4th + 1.2h^2 - 6.1t - 6.1h + 39.5\right)$

   $p'(t) = \lim\limits_{h \to 0} \dfrac{(\cancel{1.2t^2} + 2.4th + 1.2h^2 - \cancel{6.1t} - 6.1h + \cancel{39.5}) - (\cancel{1.2t^2} - \cancel{6.1t} + \cancel{39.5})}{h}$

   $= \lim\limits_{h \to 0} \dfrac{2.4th + 1.2h^2 - 6.1h}{h}$ $\qquad \left(\text{or }\ \lim\limits_{h \to 0} \dfrac{h(2.4t + 1.2h - 6.1)}{h}\right)$

   $= \lim\limits_{h \to 0} (2.4t + 1.2h - 6.1)$ $\qquad$ by the Replacement Rule

   $= 2.4t - 6.1$

   $\boxed{p'(t) = 2.4t - 6.1 \text{ dollars per year}}$

   b. $p'(3.5) = \boxed{2.3 \text{ dollars per year}}$

15. a. $f(3) = \boxed{1.469 \text{ billion gallons}}$

   b. typical point: $\left(t,\ -0.009t^2 + 0.12t + 1.19\right)$

   close point: $\left(t+h,\ -0.009(t+h)^2 + 0.12(t+h) + 1.19\right)$
   $= \left(t+h,\ -0.009t^2 - 0.018th - 0.009h^2 + 0.12t + 0.12h + 1.19\right)$

   $f'(t) = \lim\limits_{h \to 0} \dfrac{(\cancel{-0.009t^2} - 0.018th - 0.009h^2 + \cancel{0.12t} + 0.12h + \cancel{1.19}) - (\cancel{-0.009t^2} + \cancel{0.12t} + \cancel{1.19})}{h}$

   $= \lim\limits_{h \to 0} \dfrac{-0.018th - 0.009h^2 + 0.12h}{h}$ $\qquad \left(\text{or }\ \lim\limits_{h \to 0} \dfrac{h(-0.018t - 0.009h + 0.12)}{h}\right)$

   $= \lim\limits_{h \to 0} (-0.018t - 0.009h + 0.12)$ $\qquad$ by the Replacement Rule

   $= -0.018t + 0.12$

   $\boxed{f'(t) = -0.018t + 0.12 \text{ billion gallons per year}}$

   c. $f'(3) = \boxed{0.066 \text{ billion gallons per year}}$

   $\boxed{\text{In 2007, the amount of fuel used by Southwest Airlines was increasing by 66 million gallons per year.}}$

17. (The activity specifies to round the coefficients in the model to two decimal places.)
   a. $c(x) \approx 32.97x + 328.10$ gives the CPI (for all urban consumers) for college tuition and fees between 2000 and 2008 where x is the number of years since 2000, data from 2000 through 2008. CPI stands for Consumer Price Index and as all indices is a unit-less.
   b. typical point: $(x,\ 32.97x + 328.10)$
      close point: $(x+h,\ 32.97(x+h) + 328.10) = (x+h,\ 32.97x + 32.97h + 328.10)$

$$c'(x) \approx \lim_{h \to 0} \frac{(32.97x + 32.97h + 328.10) - (32.97x + 328.10)}{h}$$

$$= \lim_{h \to 0} \frac{32.97h}{h}$$

$$= \lim_{h \to 0} 32.97 \qquad \text{by the Replacement Rule}$$

$$= 32.97$$

$$\boxed{c'(x) \approx 32.97 \text{ per year}}$$

(The coefficients of the model were rounded in order to make the algebra easier. When calculating results, unrounded coefficients should still be used.)

c. $c'(5) \approx \boxed{32.973 \text{ per year}}$

$\boxed{\text{In 2005, the CPI for college tuition and fees was increasing by 32.973 per year.}}$

d. $\%\text{ROC} = \dfrac{c'(5)}{c(5)} \cdot 100\% \approx \boxed{6.689\% \text{ per year}}$

$\boxed{\text{In 2005, the CPI for college tuition and fees was increasing by 6.689\% per year.}}$

19. Answers will vary but might include the following ideas:
    Finding the rate of change graphically gives a quick estimate. Finding the rate of change algebraically can be tedious but results in a formula that can be used to find rates of change at any input point for the function. The discussion of algebraically determining rate-of-change formulas was limited to constant, linear, and quadratic equations (with a few exceptions such as activities 5 and 6 in Section 2.5).
    The numerical method was used to find the rate of change for any input value where the function was differentiable.

Section 2.6 Rate-of-Change Graphs (pages 180–184)

1. a. $\boxed{\text{zero slope: } x = a}$ where f has a minimum
        $\boxed{\text{no relative maximum or minimum slope values}}$ on the interval shown
    b. $\boxed{\text{negative slope: } x < a}$ because f is decreasing
        $\boxed{\text{positive slope: } x > a}$ because f is increasing
        $\boxed{\text{increasing slope over entire interval shown}}$:
        When $x < a$, f has negative slopes that are becoming less steep as x increases toward a.
            This means, that f has slopes that are negative but increasing toward zero for $x < a$.
        When $x > a$, f has positive slopes that are becoming more steep as x increases away from a.
            That is, f has slopes that are increasing away from zero for $x > a$.

c. One possible slope graph:
$f'(x)$

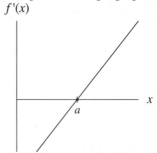

3. a. no zero slope appears on the interval shown. However, the slopes seem to be approaching zero as x decreases toward the left end of the input interval.
no relative maximum or minimum slope values on the interval shown
   b. positive and increasing slope for entire interval shown because g is increasing and is becoming more steep as x increases.
   c. One possible slope graph:
$g'(x)$

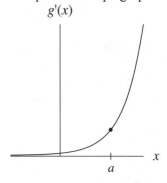

5. a. zero slope: $x = b$, $x = d$, and $x = f$
   $x = b$ where h has a maximum
   $x = d$ where h has a minimum
   $x = f$ where h has a maximum
   relative minimum slope: $x = c$ where h has an inflection point and is steeper than at surrounding points
   relative maximum slope: $x = e$ where h has an inflection point and is steeper than at surrounding points
   b. positive slope: $x < b$ and $d < x < f$ because h is increasing
   negative slope: $b < x < d$ and $x > f$ because h is decreasing
   decreasing slope: $x < c$ and $x > e$
   $x < b$ because slopes are positive and becoming closer to zero (tangent lines are becoming less steep)
   $b < x < c$ because slopes are negative and getting farther from zero (tangent lines are becoming more steep)
   $e < x < f$ because slopes are positive and becoming closer to zero (tangent lines are becoming less steep)
   $f < x$ because slopes are negative and getting farther from zero (tangent lines are becoming more steep)

2.6 Rate-of-Change Graph

Solutions to Odd-Numbered Activities

increasing slope: $c < x < e$
$c < x < d$ because slopes are negative and becoming closer to zero (tangent lines are becoming less steep)
$d < x < e$ because slopes are positive and getting farther from zero (tangent lines are becoming more steep)

c. One possible slope graph:
$h'(x)$

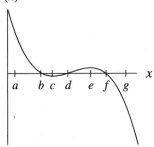

7. a. no zero slope appears on the interval shown. However, the slopes seem to be approaching zero as x increases.
no relative maximum or minimum slope values on the interval shown
b. negative and increasing slope for entire interval shown because j is decreasing and is becoming less steep (slopes are closer to zero) as x increases.
c. One possible slope graph:
$j'(x)$

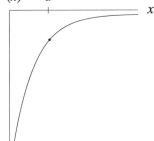

9. a. zero slope: $x = a$ where k has an inflection point
relative maximum slope: $x = a$ where k has an inflection point and is less steep than at surrounding points
b. negative slope for all x shown (except for $x = a$) because k is decreasing
c. One possible slope graph:
$k'(x)$

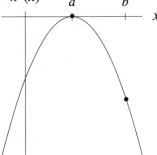

11. (The figure in the answer key in the text erroneously lists the rise for the triangle at $x = 7$ as 4 instead of 0.4.)

    a.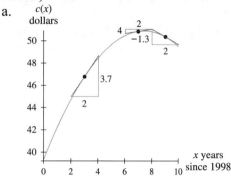

    Using rise and run estimated for tangent lines drawn on the graph,

    2001: $c'(3) \approx \dfrac{3.7}{2} = \boxed{1.85 \text{ dollars per year}}$

    2005: $c'(7) \approx \dfrac{0.4}{2} = \boxed{0.2 \text{ dollar per year}}$

    2007: $c'(9) \approx \dfrac{-1.3}{2} = \boxed{-0.65 \text{ dollars per year}}$

    b.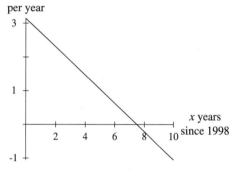

13. a.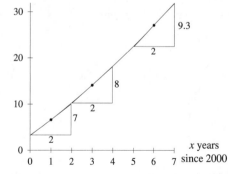

    Using rise and run estimated for tangent lines drawn on the graph,

    2001: $f'(1) \approx \dfrac{7}{2} = \boxed{3.5 \text{ hundred thousand cases per year}}$

2003: $f'(3) \approx \dfrac{8}{2} =$ 4 hundred thousand cases per year   (The answer key in the book erroneously reports the year 2003 as 2002.)

2006: $f'(6) \approx \dfrac{9.3}{2} =$ 4.65 hundred thousand cases per year (The answer key in the book rounds this result.)

b.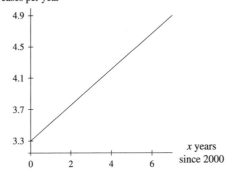

15. a.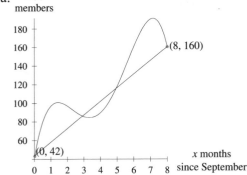

Using points (0, 42) and (8, 160) estimated from the graph,

$$\text{AROC} = \dfrac{160 - 42}{8 - 0} = \boxed{14.75 \text{ members per month}}$$

b.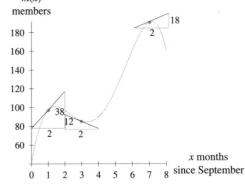

Using rise and run estimated for tangent lines drawn on the graph,

October: $m'(1) \approx \dfrac{38}{2} =$ 19 members per month

December: $m'(3) \approx \dfrac{-12}{2} = \boxed{-6 \text{ members per month}}$

April: $m'(7) \approx \dfrac{18}{2} = \boxed{9 \text{ members per month}}$

c. (The vertical axis label in the answer key in the book is missing the prime in $m'(x)$.)

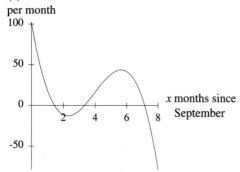

17. a. The derivative does not exist at $t = 5$ because the limit of secant lines from the right of that point does not equal the limit of secant lines from the left of that point.

b.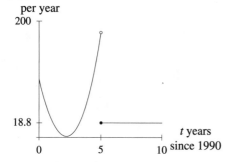

19. a.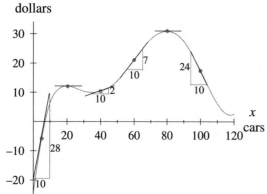

Using rise and run estimated for tangent lines drawn on the graph,

5 cars: $p'(5) \approx \dfrac{28}{10} = \boxed{2.8 \text{ thousand dollars per car}}$

20 cars: $p'(20) \approx \dfrac{0}{10} = \boxed{0 \text{ thousand dollars per car}}$

40 cars: $p'(40) \approx \dfrac{2}{10} = \boxed{0.2 \text{ thousand dollars per car}}$

60 cars: $p'(60) \approx \dfrac{7}{10} = \boxed{0.7 \text{ thousand dollars per car}}$

80 cars: $p'(80) \approx \dfrac{0}{10} = \boxed{0 \text{ thousand dollars per car}}$

100 cars: $p'(100) \approx \dfrac{-12}{10} = \boxed{-1.2 \text{ thousand dollars per car}}$

b. inflection points near $\boxed{28 \text{ cars, } 60 \text{ cars, and } 100 \text{ cars}}$

c.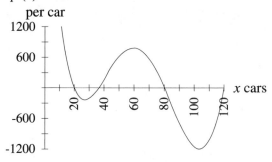

21. a. There is no derivative at $x = 0$ because y has a vertical asymptote at this input value, so is not continuous.

    There is no derivative at either $x = 3$ or $x = 4$ because the left- and right-hand limits of output values are not equal at these input values, so the function is not continuous.

    b. (The figure in the answer key in the book should have open circles instead of closed circles at $x = 3$ and $x = 5$.)
    One possible slope graph:

    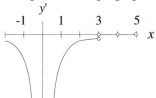

23. a. There is no derivative at $x = 2$ or at $x = 3$ because even though the function is continuous at these input values, at neither point does the limit of secants from the left equal the limit of secants from the right.

b. One possible slope graph:

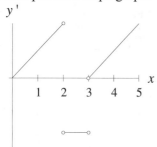

25. (The answer key in the graph erroneously shows the discontinuity in the slope graph occurring at $t = 3$ instead of $t = 2$.)
One possible slope graph:

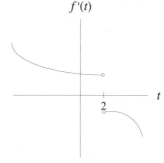

27. Answers will vary but might include the following idea:
    The horizontal-axis intercepts of the rate-of-change graph occur at the maxima, minima, or inflection points of the function.

Chapter 2 Review Activities (pages 185–190)

1. (The answer key in the text erroneously reports -5.293 instead of -5.393 for the both change and the AROC.)

    change = new − old

    $\quad\quad\quad = 92.772 - 98.165$

    $\quad\quad\quad = \boxed{-5.393 \text{ million enplaned passengers}}$

    $\boxed{\text{The number of paying passengers on American Airlines decreased by 5.393 million between 2007 and 2008.}}$

    percentage change $= \dfrac{\text{change}}{\text{old}} \cdot 100\%$

    $\quad\quad\quad\quad\quad\quad\quad = \dfrac{-5.393}{98.165} \cdot 100\%$

    $\quad\quad\quad\quad\quad\quad\quad \approx \boxed{-5.49\%}$

    $\boxed{\text{The number of paying passengers on American Airlines decreased by 5.49\% between 2007 and 2008.}}$

$$\text{AROC} = \frac{\text{change}}{\text{length of interval}}$$

$$= \frac{-5.393}{2008 - 2007}$$

$$= \boxed{-5.393 \text{ million enplaned passengers per year}}$$

The number of paying passengers on American Airlines decreased by an average of 5.393 million per year between 2007 and 2008.

3. Using points (2000, 1), (2008, 120), and (2011, 125) estimated from the graph,
   a. change = new − old

   $$= 120 - 1$$

   $$= \boxed{119 \text{ million cameras}}$$

   Digital still camera sales increased by 119 million cameras between 2000 and 2008.

   b. $\text{AROC} = \dfrac{\text{change}}{\text{length of interval}}$

   $$= \frac{119}{2008 - 2000}$$

   $$= \boxed{14.875 \text{ million cameras per year}}$$

   Digital still camera sales increased by an average of 14.875 million cameras per year between 2000 and 2008.

   c. $\text{percentage change} = \dfrac{\text{new} - \text{old}}{\text{old}} \cdot 100\%$

   $$= \frac{125 - 120}{120} \cdot 100\%$$

   $$\approx \boxed{4.167\%}$$

   Digital still camera sales are projected to increase by approximately 4.167% between 2008 and 2011.

5. a. $S'(t)$ is greatest at point B because all three slopes are non-negative and the tangent line at B is steeper than tangent lines at A or C.

   b.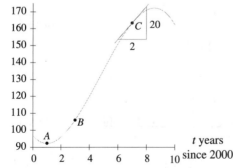

Using rise and run estimated from a tangent line drawn on the graph,

slope = $\frac{20}{2}$ = 10 billion dollars per year

c. $S'(7) \approx 10$

d. In 2007, annual U.S. factory sales of consumer electronics goods to dealers were increasing by $10 billion per year.

7. a. A, E, D, C, B, F
   Steepness does not depend on whether the slope is positive or negative.

   b. A: concave up
      B: neither
      C: concave down
      D: neither
      E: concave up
      F: concave up

   c. A: lie below
      B: cut through
      C: lie above
      D: cut through
      E: lie below
      F: lie below

   d.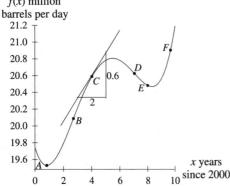

   ROC = $\frac{0.6}{2}$ = 0.3 million barrels per day

   In 2004, U.S. annual oil consumption was increasing by approximately 300,000 barrels per day.

9. a. units of measure for output of $\frac{df}{dt} = \frac{\text{output units of } f}{\text{input unit of } f}$

   $= \frac{\text{pounds}}{\text{minute}}$

   $=$ pounds per minute

   b. no; While the grill is on, the propane in the tank is decreasing and so the rate of change must be negative.

   c. When the grill has been on for 10 minutes, the propane in the tank is decreasing by 0.23 pounds per minute.

11. One possible slope graph:

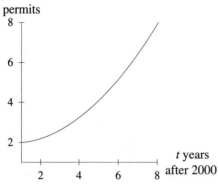

13. a.

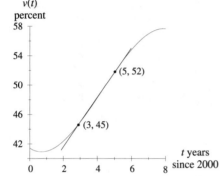

Using points $(5, 52)$ and $(3, 45)$ estimated from a tangent line drawn on the graph,

$$v'(5) \approx \frac{52-45}{5-3} = \boxed{3.5 \text{ percentage points per year}}$$

b. $\%\text{ROC} = \frac{3.5}{52} \cdot 100\% \approx \boxed{6.731 \% \text{ per year}}$

15. a.

$x \to 40^-$	Slope of secant $= \dfrac{w(40)-w(x)}{40-x}$	$x \to 40^+$	Slope of secant $= \dfrac{w(40)-w(x)}{40-x}$
39.9	$\dfrac{m(40)-m(39.9)}{40-39.9} \approx 0.5621$	40.1	$\dfrac{m(40)-m(40.1)}{40-40.1} \approx 0.5579$
39.99	$\dfrac{m(40)-m(39.99)}{40-39.99} \approx 0.56021$	40.01	$\dfrac{m(40)-m(40.01)}{40-40.01} \approx 0.55979$
39.999	$\dfrac{m(40)-m(39.999)}{40-39.999} \approx 0.560021$	40.001	$\dfrac{m(40)-m(40.001)}{40-40.001} \approx 0.559979$
39.9999	$\dfrac{m(40)-m(39.9999)}{40-39.9999} \approx 0.5600021$	40.0001	$\dfrac{m(40)-m(40.0001)}{40-40.0001} \approx 0.5599979$
	$\lim\limits_{x \to 40^-} \left(\begin{array}{c}\text{slopes of}\\\text{secants}\end{array}\right) \approx 0.56$		$\lim\limits_{x \to 40^+} \left(\begin{array}{c}\text{slopes of}\\\text{secants}\end{array}\right) \approx 0.56$

$$w'(40) = \lim_{x \to 40} \left(\text{slopes of secants}\right) = \lim_{x \to 40} \frac{w(40)-w(x)}{40-x} \approx \boxed{0.56 \text{ pounds per year (of age)}}$$

b. $\%ROC = \dfrac{w'(40)}{w(40)} \cdot 100\% \approx \boxed{0.3\% \text{ per year}}$

17. a. Using the cubic part of the piecewise model,
    $L(8) = 27.67 \rightarrow \boxed{28 \text{ launches}}$ (Round to the nearest whole number.)
    b. Using the cubic part of the piecewise model,

$t \to 8^-$	Slope of secant $= \dfrac{L(8)-L(t)}{8-t}$	$t \to 8^+$	Slope of secant $= \dfrac{L(8)-L(t)}{8-t}$
7.9	$\dfrac{L(8)-L(7.9)}{8-7.9} \approx 5.87$	8.1	$\dfrac{L(8)-L(8.01)}{8-8.01} \approx 6.21$
7.99	$\dfrac{L(8)-L(7.99)}{8-7.99} \approx 6.02$	8.01	$\dfrac{L(8)-L(8.01)}{8-8.01} \approx 6.06$
7.999	$\dfrac{L(8)-L(7.999)}{8-7.999} \approx 6.04$	8.001	$\dfrac{L(8)-L(8.001)}{8-8.001} \approx 6.04$
7.9999	$\dfrac{L(8)-L(7.9999)}{8-7.9999} \approx 6.04$	8.0001	$\dfrac{L(8)-L(8.0001)}{8-8.0001} \approx 6.04$
7.99999	$\dfrac{L(8)-L(7.99999)}{8-7.99999} \approx 6.04$	8.00001	$\dfrac{L(8)-L(8.00001)}{8-8.00001} \approx 6.04$
	$\lim\limits_{t \to 8^-}\left(\text{slopes of secants}\right) \approx 6$		$\lim\limits_{t \to 8^+}\left(\text{slopes of secants}\right) \approx 6$

$L'(8) = \lim\limits_{t \to 8}(\text{slopes of secants}) = \lim\limits_{t \to 8}\dfrac{L(8)-L(t)}{8-t} \approx \boxed{6 \text{ launches per year}}$

19. a. typical point: $(x,\ 7.2x^2)$

    close point: $\left(x+h,\ 7.2(x+h)^2\right) = \left(x+h,\ 7.2x^2 + 14.4xh + 7.2h^2\right)$

    $f'(x) = \lim\limits_{h \to 0} \dfrac{\left(\cancel{7.2x^2} + 14.4xh + 7.2h^2\right) - \left(\cancel{7.2x^2}\right)}{h}$

    $= \lim\limits_{h \to 0} \dfrac{14.4xh + 7.2h^2}{h}$ $\qquad\left(\text{or } \lim\limits_{h \to 0}\dfrac{h(14.4x+7.2h)}{h}\right)$

    $= \lim\limits_{h \to 0}(14.4x + 7.2h)$ $\qquad$ by the Replacement Rule

    $= 14.4x$

    $\boxed{f'(x) = 14.4x}$

    b. $f'(-2) = \boxed{-28.8}$

21. a. $m(x) = 15.32x - 63.36$ million users gives the number of mobile Internet users where x is the number of years since 2000, based on data from 2008 and projections to 2013.
    b. typical point: $(x,\ 15.32x - 63.36)$
    close point: $\left(x+h,\ 15.32(x+h) - 63.36\right) = (x+h,\ 15.32x + 15.32h - 63.36)$

$$m'(x) = \lim_{h \to 0} \frac{(15.32x + 15.32h - 63.36) - (15.32x - 63.36)}{h}$$

$$= \lim_{h \to 0} \frac{15.32h}{h}$$

$$= \lim_{h \to 0} 15.32 \qquad \text{by the Replacement Rule}$$

$$= 15.32$$

$m'(x) = 15.32$ million users per year

c. $m'(11) = \boxed{15.32 \text{ million users per year}}$

In 2011, the number of mobile Internet users was increasing by 15.32 million users per year.

23. a. no zero slope on the interval shown
no relative maximum or minimum slope values on the interval shown

b. negative slope over entire interval shown because y is decreasing
constant slope over entire interval shown : The function is linear so has constant rate of change.

c. One possible slope graph:

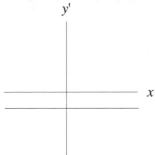

25. a. zero slope: $x = 0$, $x = b$, and $x = d$ where y has either a relative minimum or maximum
relative maximum slope: $x = a$ where y has an inflection point and is steeper than at surrounding points
relative minimum slope: $x = c$ where y has an inflection point and is steeper than at surrounding points and the slope is negative

b. negative slope: $x < 0$ and $b < x < d$ because y is decreasing
positive slope: $0 < x < b$ and $x > d$ because y is increasing
increasing slope: $x < a$ and $c < x$

$x < 0$ because slopes are negative and becoming closer to zero (tangent lines are becoming less steep)

$0 < x < a$ because slopes are positive and getting farther from zero (tangent lines are becoming more steep)

$c < x < d$ because slopes are negative and becoming closer to zero (tangent lines are becoming less steep)

$x > d$ because slopes are positive and getting farther from zero (tangent lines are becoming more steep)

decreasing slope: $a < x < c$

$a < x < b$ because slopes are positive and becoming closer to zero (tangent lines are becoming less steep)

$b < x < c$ because slopes are negative and getting farther from zero (tangent lines are becoming more steep)

c. One possible slope graph:

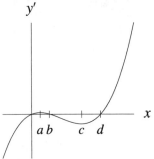

27. a.

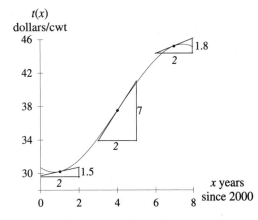

Using rise and run estimated for tangent lines drawn on the graph,

$$t'(1) \approx \frac{1.5}{2} = \boxed{\$0.75 \text{ per cwt/year}}$$

$$t'(4) \approx \frac{7}{2} = \boxed{\$3.50 \text{ per cwt/year}}$$

$$t'(7) \approx \frac{1.8}{2} = \boxed{\$0.90 \text{ per cwt/year}}$$

b. (The graph in the answer key does not match the information in part *a*.)
One possible slope graph:

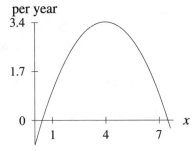

29. a. The slope of s is not defined for $x = 5$ because the function is not continuous at that value.
   b. One possible slope graph:

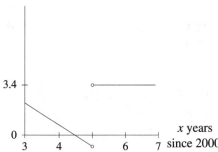

Chapter 3 Determining Change: Derivatives

Section 3.1 Simple Rate-of-Change Formulas (pages 198–200)

1. $y = 17.5$
   $\boxed{y' = 0}$ by the Constant Rule

3. $f(g) = 24\pi$
   $\boxed{\dfrac{df}{dg} = 0}$ by the Constant Rule

5. $f(x) = x^5$
   $\dfrac{df}{dx} = 5x^{5-1}$ by the Power Rule
   $\boxed{\dfrac{df}{dx} = 5x^4}$

7. $y = x^{-0.7}$
   $y' = -0.7x^{-0.7-1}$ by the Power Rule
   $\boxed{y' = -0.7x^{-1.7}}$

9. $x(t) = t^{2\pi}$
   $\boxed{x'(t) = 2\pi t^{2\pi - 1}}$ by the Power Rule

11. $y = 23x^7$
    $\dfrac{dy}{dx} = 23\left(7x^{7-1}\right)$ by the Constant Multiplier and Power Rules
    $\boxed{\dfrac{dy}{dx} = 161x^6}$

13. $f(x) = -0.5x^2$
    $f'(x) = -0.5\left(2x^{2-1}\right)$ by the Constant Multiplier and Power Rules
    $\boxed{f'(x) = -1.0x}$

15. $y = 12x^4 + 13x^3 + 5$
    $y' = 12\left(4x^{4-1}\right) + 13\left(3x^{3-1}\right) + 0$ by the Sum, Constant Multiplier, Power, and Constant Rules
    $\boxed{y' = 48x^3 + 39x^2}$

17. $y = 5x^3 + 3x^2 - 2x - 5$

$\frac{dy}{dx} = 5(3x^{3-1}) + 3(2x^{2-1}) - 2(x^{1-1}) - 0$ by the Sum, Constant Multiplier, Power, and Constant Rules

$\boxed{\frac{dy}{dx} = 15x^2 + 6x - 2}$

19. $f(x) = \frac{7}{x^3}$ rewrite as $f(x) = 7x^{-3}$

$f'(x) = 7(-3x^{-3-1})$ by the Constant Multiplier and Power Rules

$\boxed{f'(x) = -21x^{-4}}$

21. $y = \frac{-9}{x^2}$ rewrite as $y = -9x^{-2}$

$y' = -9(-2x^{-2-1})$ by the Constant Multiplier and Power Rules

$\boxed{y' = 18x^{-3}}$

23. $f(x) = 4\sqrt{x} + 3.3x^3$ rewrite as $f(x) = 4x^{0.5} + 3.3x^3$

$f'(x) = 4(0.5x^{0.5-1}) + 3.3(3x^{3-1})$ by the Constant Multiplier and Power Rules

$\boxed{f'(x) = 2x^{-0.5} + 9.9x^2}$

25. $j(x) = \frac{3x^2 + 1}{x}$ rewrite as $j(x) = 3x + x^{-1}$

$\frac{dj}{dx} = 3x^{1-1} + (-1x^{-1-1})$ by Sum, Constant Multiplier, and Power Rules

$\boxed{\frac{dj}{dx} = 3 - x^{-2}}$

27. a. $f'(x) = 0.012x^2 - 0.122x + 0.299$ dollars per year gives the rate of change in the average charge to non-account holders who use an ATM, where x is the number of years after 1998, based on data between 1998 and 2008.
    b. $f(13) \approx \$3.265$ → $\boxed{\$3.27}$ (Round to the nearest cent.)
    c. $f'(11) \approx \boxed{0.409 \text{ dollars per year}}$ (Rates do not need to be rounded to the nearest cent.)

29. a. $t'(x) = -1.6x + 11.6$ °F per hour
    b. $t'(4) = \boxed{5.2°F \text{ per hour}}$
    c. $t'(10) = \boxed{-4.4 °F \text{ per hour}}$

d.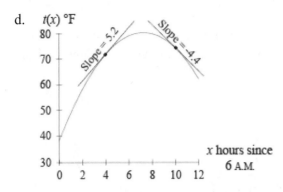

31. a. 2011: $n(11) \approx \boxed{42.481 \text{ million people}}$
    2030: $n(30) = \boxed{70.87 \text{ million people}}$

   b. $n'(x) = -0.00246x^2 + 0.118x + 0.183$ million people per year
   2011: $n'(11) \approx \boxed{1.183 \text{ million people per year}}$
   2030: $n'(30) = \boxed{1.509 \text{ million people per year}}$

   c. $\dfrac{n'(30)}{n(30)} \approx 0.0213 = \boxed{2.13 \text{ percent per year}}$

33. a. $m(x) \approx 6.930x + 682.188$ kilocalories/day gives the metabolic rate of a typical 18- to 30-year-old male who weighs x pounds, based on weights from 88 to 200 pounds.
   b. $m'(x) \approx 6.930$ kilocalories/day per pound gives the rate of change in the metabolic rate of a typical 18- to 30-year-old male who weighs x pounds, data from 88 to 200 pounds.
   c. The metabolic rate of a typical 18- to 30- year old male is increasing by approximately 6.930 kilocalories/day per pound regardless of body-weight.

35. a. revenue: $R(n) \approx -29.330n^2 + 1450.393n - 3000.777$ dollars gives the revenue from n pageant gowns, data from $3 \le n \le 15$.
   cost: $C(n) \approx -2.158n^2 + 152.143n - 199.271$ dollars gives the cost to make n pageant gowns, data from $3 \le n \le 15$.
   profit: $P(n) = R(n) - C(n) \approx -27.173n^2 + 1298.25n - 2801.506$ dollars gives the profit from n pageant gowns, data from $3 \le n \le 15$. (The coefficient for n^2 is −27.173 instead of −27.172 because the unrounded coefficients for functions R and C are used.)
   ROC: $P'(n) \approx -54.345n + 1298.25$ dollars per gown gives the profit from n pageant gowns, data from $3 \le n \le 15$.

   b. 2 gowns: $P'(2) \approx \boxed{1189.560 \text{ dollars per gown}}$
   $\boxed{\text{When the seamstress makes two gowns, her profit is increasing by \$1189.56 per gown.}}$
   10 gowns: $P'(10) \approx \boxed{754.798 \text{ dollars per gown}}$
   $\boxed{\text{When the seamstress makes ten gowns, her profit is increasing by \$754.80 per gown.}}$

   c. (The answer key in the book uses $\overline{A}$ instead of the preferred notation $\overline{C}$.)
   average cost: $\overline{C}(n) = \dfrac{C(n)}{n} \approx \dfrac{-2.158n^2 + 152.143n - 199.271}{n}$ dollars gives the average cost to make a pageant gown when n gowns are made, data from $3 \le n \le 15$.
   Alternate form: $\overline{C}(n) \approx -2.158n + 152.143 - \dfrac{199.271}{n}$

3.1 Simple Rate-of-Change Formulas
Solutions to Odd-Numbered Activities

ROC: $\dfrac{d\overline{C}}{dn} = -2.158 + \dfrac{199.271}{n^2}$ dollars per gown gives the rate of change in the average cost to make a pageant gown when n gowns are made, data from $3 \le n \le 15$.

d. 2 gowns: $\left.\dfrac{d\overline{C}}{dn}\right|_{n=2} \approx$ 47.660 dollars per gown

  6 gowns: $\left.\dfrac{d\overline{C}}{dn}\right|_{n=6} \approx$ 3.378 dollars per gown

  12 gowns: $\left.\dfrac{d\overline{C}}{dn}\right|_{n=12} \approx$ −0.774 dollars per gown

37. a. $P(x) = 175 - C(x) = 175 - \left(0.015x^2 - 0.78x + 46 + \dfrac{49.6}{x}\right)$ dollars

    Alternate form: $P(x) = 129 - 0.015x^2 + 0.78x - \dfrac{49.6}{x}$

  b. $P'(x) = -0.030x + 0.78 + \dfrac{49.6}{x^2}$ dollars per window

  c. $P(80) =$ $94.78

  d. $P'(80) \approx$ −$1.612 per window

   When Windowlux produces 80 windows per hour, their profit per window is decreasing by approximately $1.61 per (additional) window produced.

39. Answers will vary but should be similar to the following:
    The following figures show the two possible graphs of cubic functions of the form
    $f(x) = ax^3 + bx^2 + cx + d$ where $a > 0$.

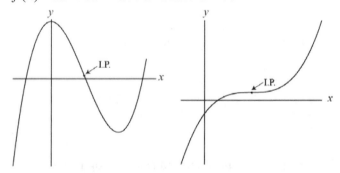

    If $a > 0$, the slopes of the graph of $f(x) = ax^3 + bx^2 + cx + d$ are decreasing as x increases until x reaches the input value associated with the inflection point of f. As x increases beyond the input value of the inflection point, the slopes of the graph of $f(x) = ax^3 + bx^2 + cx + d$ are increasing. The farther (in either direction) x is from the input value of the inflection point, the greater the magnitude of the slope of f. The description of the behavior of the slopes of a cubic function also describes the behavior of the output of a concave up quadratic function (a parabola).

Section 3.2 Exponential, Logarithmic, and Cyclic Rate-of-Change Formulas (pages 209–211)

1. $h(x) = 3 - 7e^x$
   $\boxed{h'(x) = -7e^x}$ by the Difference, Constant, Constant Multiple, and e^x Rules

3. $g(x) = 2.1^x + \pi^2$
   $\boxed{g'(x) = (\ln 2.1) \cdot 2.1^x}$ by the Sum, Exponential, and Constant Rules

5. $h(x) = 12(1.6^x)$
   $\boxed{h'(x) = 12(\ln 1.6) \cdot 1.6^x}$ by the Constant Multiplier and Exponential Rules

7. $f(x) = 10\left(1 + \dfrac{0.05}{4}\right)^{4x}$ rewrite as $f(x) \approx 10(1.05095)^x$
   $\boxed{f'(x) \approx 10(\ln 1.05095) \cdot 1.05095^x}$ by the Constant Multiplier and Exponential Rules
   Alternate method: rewrite f as $f(x) = 10(1.0125^4)^x$
   $f'(x) = 10 \cdot \ln(1.0125^4) \cdot 1.0125^{4x}$

9. $j(x) = 4.2(0.8^x) + 3.5$
   $\boxed{\dfrac{dj}{dx} = 4.2(\ln 0.8) \cdot 0.8^x}$ by the Sum, Constant Multiplier, Exponential, and Constant Rules

11. $j(x) = 4\ln x - e^\pi$
    $\dfrac{dj}{dx} = 4\dfrac{1}{x}$ by the Difference, Constant Multiplier, Natural Log, and Constant Rules
    $\boxed{\dfrac{dj}{dx} = \dfrac{4}{x}}$

13. $g(x) = 12 - 7\ln x$
    $\dfrac{dg}{dx} = -7\dfrac{1}{x}$ by the Difference, Constant, Constant Multiplier, and Natural Log Rules
    $\boxed{\dfrac{dg}{dx} = \dfrac{-7}{x}}$

15. $n = 14\sin x$
    $\boxed{\dfrac{dn}{dx} = 14\cos x}$ by the Constant Multiplier and Sine Rules

17. $g(x) = 6\ln x - 13\sin x$

$\dfrac{dg}{dx} = 6\dfrac{1}{x} - 13\cos x$ by the Difference, Constant Multiplier, Natural Log, and Sine Rules

$\boxed{\dfrac{dg}{dx} = \dfrac{6}{x} - 13\cos x}$

19. $f(t) = 0.07\cos t - 4.7\sin t$

$f'(t) = 0.07(-\sin t) - 4.7\cos t$ by the Difference, Constant Multiplier, Cosine, and Sine Rules

$\boxed{f'(t) = -0.07\sin t - 4.7\cos t}$

21. a. $f(t) = 1000e^{0.07t}$ rewrite as $f(t) \approx 1000(1.0725^t)$

$\boxed{f'(t) \approx 1000(\ln 1.0725) \cdot 1.0725^t \text{ dollars per year}}$

b. $f'(10) \approx \boxed{\$140.963 \text{ per year}}$
(Result is calculated using unrounded numbers in calculations.)

23. a. weight: $w(7) \approx \boxed{25.641 \text{ grams}}$

ROC: $w'(t) = \dfrac{7.37}{t}$ grams per week

$w'(7) \approx \boxed{1.053 \text{ grams per week}}$

b. $\dfrac{w(9) - w(5)}{9 - 5} \approx \boxed{1.083 \text{ grams per week}}$

c. Answers will vary but should refer to the derivative function:
The rate at which the mouse is growing decreases as the mouse gets older. The rate of change in the mouse's weight is given by the formula $w'(t) = \dfrac{7.37}{t}$. As t increases, $w'(t)$ will decrease.

25. a. $p'(t) = \begin{cases} -23.73t^2 + 242t + 194 & \text{when } 0.7 < t < 13 \\ 45{,}500(\ln 0.847) \cdot 0.847^t & \text{when } 13 < t < 55 \end{cases}$

where output is measured in people per year gives the rate of change of the population of Aurora, Nevada t years since 1859.

b. 1870: $p'(11) = \boxed{-15.33 \text{ people per year}}$ (Use the $0.7 < t < 13$ piece of the derivative function.)
1900: $p'(41) \approx \boxed{-8.346 \text{ people per year}}$ (Use the $13 < t < 55$ piece of the derivative function.)

27. a. $F(t) = 1000e^{0.043t}$ dollars gives the value of a $1000 investment after t years at 4.3% compounded continuously.

b. Rewrite F as $F(t) \approx 1000(1.044^t)$

$\boxed{F'(t) \approx 1000(\ln 1.044) \cdot 1.044^t \text{ dollars per year}}$

Alternate method: Rewrite F as $F(t) = 1000(e^{0.043})^t$

$F'(t) = 1000(\ln e^{0.043}) \cdot (e^{0.043})^t = 43e^{0.043t}$

c. $F(5) \approx \$1239.862 \rightarrow \boxed{\$1239.86}$ (Round to the nearest cent.)
d. $F'(5) \approx \boxed{\$53.314 \text{ per year}}$ (Rates do not need to be rounded to the nearest cent.)

29. a. Answers will vary but should include the following information:
   A scatter plot suggests a concave down, increasing function with output values that do not approach a specific limit but continue to increase more and more slowly.
   b. $f(t) \approx 100.527 + 11.912 \ln t$ million homes gives the number of homes in the United States eligible for a high-speed Internet connection, in year $t + 2002.5$, based on data from 2003 $(t = 0.5)$ to 2008 $(t = 5.5)$.
   c. $f'(t) \approx \dfrac{11.912}{t}$ million homes per year gives the rate of change in the number of homes in the United States eligible for a high-speed Internet connection, in year $t + 2002.5$, based on data from 2003 $(t = 0.5)$ to 2008 $(t = 5.5)$.
   d. 2005 homes: $f(2.5) \approx \boxed{111.442 \text{ million homes}}$
   ROC: $f'(2.5) \approx \boxed{4.765 \text{ million homes per year}}$
   %ROC: $\dfrac{f'(2.5)}{f(2.5)} \approx 0.043 = \boxed{4.3\% \text{ per year}}$
   In 2005, 111,442,000 homes in the United States were eligible for a high-speed Internet connection and the number of eligible homes was increasing by approximately 4,765,000 homes (or 4.3%) per year.
   2010 homes: $f(7.5) \approx \boxed{124.530 \text{ million homes}}$
   ROC: $f'(7.5) \approx \boxed{1.588 \text{ million homes per year}}$
   %ROC: $\dfrac{f'(7.5)}{f(7.5)} \approx 0.013 = \boxed{1.3\% \text{ per year}}$
   In 2010, 124,530,000 homes in the United States were eligible for a high-speed Internet connection and the number of eligible homes was increasing by approximately 1,588,000 homes (or 1.3%) per year.

31. Answers will vary but should be similar to the following:
   A function of the form $g(x) = e^{kx}$ where $k \neq 0$ can be rewritten as $g(x) = (e^k)^x$. Because e^k is a constant the Exponential Rule for derivatives can be applied, so $g'(x) = (\ln e^k) \cdot e^{kx}$. One of the properties of logarithms states that $\ln e^k = k$. So $g'(x) = ke^{kx}$.

Section 3.3 Rates of Change for Functions that can be Composed (pages 216–219)

1. a. When $t = 5$ million dollars is invested $\boxed{w(5) = 2100 \text{ workers}}$ are needed.
   $\boxed{\text{When \$5 million is invested in technology for a manufacturing plant, the plant needs 2100 workers to maximize production.}}$
   b. When 5 million dollars is invested, $w = 2100$ workers are employed with labor costs of $\boxed{L(2100) = 32 \text{ million dollars}}$.

3.3 Rates of Change for Functions that can be Composed
Solutions to Odd-Numbered Activities

When 2100 workers are employed by a manufacturing plant, labor costs for that plant are $32 million.

c. When $t = 5$ million dollars is invested the number of workers needed is increasing by $\left.\dfrac{dw}{dt}\right|_{t=5} = 200$ workers per million dollars.

When $5 million is invested in technology for a manufacturing plant, the number of workers needed to maximize production is increasing by 200 workers per million dollars.

d. (The activity in the book is missing the word *million* in the last line of the first paragraph. The activity should state, "...labor costs are increasing by approximately $0.024 million per worker.") When $w = 2100$ workers are employed labor costs are increasing by $\left.\dfrac{dL}{dw}\right|_{w=2100} = \$0.024$ million per worker.

When 2100 workers are employed by a manufacturing plant, labor costs for that plant are increasing by $0.024 million per worker.

e. The output of function w is identical to the input of function L so the two functions can be combined using composition: $L(w(t))$.

$$\dfrac{dL}{dt} = \dfrac{dL}{dw} \cdot \dfrac{dw}{dt}$$

$$\left.\dfrac{dL}{dw}\right|_{t=5} = \dfrac{0.024 \text{ million dollars (labor)}}{\text{worker}} \cdot \dfrac{200 \text{ worker}}{\text{million dollars (technology)}}$$

$$\left.\dfrac{dL}{dw}\right|_{t=5} = \$4.8 \text{ per dollar} \quad \text{(labor cost dollars per technology dollar)}$$

When $5 million is invested in technology, labor costs are increasing by $4.8 per dollar of invested.

3. a. On February 13, $n = 44$ and $v(44) = 150$ thousand pieces of mail.
   On February 13, 150 thousand pieces of mail are processed.

   b. On the 44th day of the year, $v = 150$ thousand pieces and $h(150) = 180$ hours.
   On February 13, 150 thousand pieces of mail are processed during 180 employee-hours.

   c. On the 44th day of the year, $\left.\dfrac{dv}{dn}\right|_{n=44} = -0.3$ thousand pieces per day.
   On February 13, the amount of mail being processed is decreasing by 0.3 thousand pieces per day.

   d. When $v = 150$ thousand pieces, $\left.\dfrac{dh}{dv}\right|_{v=150} = 12$ hours per thousand pieces.
   On February 13, the amount of time necessary to processes mail is increasing by 12 hours per thousand pieces.

   e. The output of function v is identical to the input of function h so the two functions can be combined using composition: $h(v(n))$.

$$\frac{dh}{dn} = \frac{dh}{dv} \cdot \frac{dv}{dn}$$

$$\left.\frac{dh}{dn}\right|_{n=44} = \frac{12 \text{ hours}}{\cancel{\text{thousand pieces}}} \cdot \frac{-0.3 \cancel{\text{ thousand pieces}}}{\text{day}}$$

$$\boxed{\left.\frac{dh}{dn}\right|_{n=44} = -3.6 \text{ hours per day}}$$

On February 13, the number of employee-hours at the post office was decreasing by 3.6 hours per day.

5. Let $g(t)$ gallons represent the amount of gasoline in the tank at time t days.
   Let $v(g)$ dollars represent the value of g gallons of gasoline.
   The output of function g is identical to the input of function v so the two functions can be combined using composition: $v(g(t))$.

$$\frac{dv}{dt} = \frac{dv}{dg} \cdot \frac{dg}{dt}$$

$$= \frac{2.51 \text{ dollars}}{\cancel{\text{gallon}}} \cdot \frac{-3.5 \cancel{\text{ gallons}}}{\text{day}}$$

$$= \boxed{-8.785 \text{ dollars per day}}$$

7. The output $t(f)$ dollars represents the average ticket price when f dollars is the operating cost.
   The constant rate of change for t is given as the slope of the line: 0.3 dollars per dollar (dollar ticket price per dollar operating cost).
   Let $f(b)$ dollars represent the operating cost when b dollars is the price of a barrel of oil.
   The rate of change for f is measured in dollars operating cost per dollar oil price.
   The output of function f is identical to the input of function t so the two functions can be combined using composition: $t(f(b))$.

$$\frac{dt}{db} = \frac{dt}{df} \cdot \frac{df}{db}$$

$$= \frac{0.3 \text{ dollars (ticket price)}}{\cancel{\text{dollar (operating cost)}}} \cdot \frac{200 \cancel{\text{ dollars (operating cost)}}}{\text{dollar (oil price)}}$$

$$= \boxed{\$60 \text{ per dollar}} \text{ (dollar ticket price per dollar oil price)}$$

9. $\boxed{c(x(t)) = 3(4-6t)^2 - 2}$

$$\frac{dc}{dt} = \frac{dc}{dx} \cdot \frac{dx}{dt}$$

$$= 6x \cdot (-6)$$

$$= 6(4-6t)(-6)$$

$$\boxed{\frac{dc}{dt} = 6(4-6t)(-6)}$$

Alternate form: $\dfrac{dc}{dt} = -36(4-6t)$
$= -144 + 216t$

11. $\boxed{h(p(t)) = \dfrac{4}{1+3e^{-0.5t}}}$

Rewrite h using a negative exponent: $h(p) = 4p^{-1}$

Rewrite p as a composite function of $p(u) = 1+3e^u$ and $u(t) = -0.5t$

$\dfrac{dh}{dt} = \dfrac{dh}{dp} \cdot \dfrac{dp}{du} \cdot \dfrac{du}{dt}$

$= -4p^{-2} \cdot 3e^u \cdot (-0.5)$

$= -4(1+3e^u)^{-2} \cdot 3e^u \cdot (-0.5)$

$= -4(1+3e^{-0.5t})^{-2} \cdot 3e^{-0.5t} \cdot (-0.5)$

$\boxed{\dfrac{dh}{dt} = -4(1+3e^{-0.5t})^{-2} \cdot 3e^{-0.5t} \cdot (-0.5)}$

Alternate form: $\dfrac{dh}{dt} = 6e^{-0.5t}(1+3e^{-0.5t})^{-2}$

$= \dfrac{6e^{-0.5t}}{(1+3e^{-0.5t})^2}$

13. $\boxed{k(t(x)) = 4.3(\ln x)^3 - 2(\ln x)^2 + 4\ln x - 12}$

$\dfrac{dk}{dx} = \dfrac{dk}{dt} \cdot \dfrac{dt}{dx}$

$= (12.9t^2 - 4t + 4) \cdot \dfrac{1}{x}$

$= (12.9(\ln x)^2 - 4\ln x + 4) \cdot \dfrac{1}{x}$

$\boxed{\dfrac{dk}{dx} = (12.9(\ln x)^2 - 4\ln x + 4) \cdot \dfrac{1}{x}}$

Alternate form: $\dfrac{dk}{dx} = \dfrac{12.9(\ln x)^2 - 4\ln x + 4}{x}$

15. $\boxed{p(t(k)) = 7.9\sin(14k^3 - 12k^2)}$

$\dfrac{dp}{dk} = \dfrac{dp}{dt} \cdot \dfrac{dt}{dk}$

$= 7.9\cos t \cdot (42k^2 - 24k)$

$= 7.9\cos(14k^3 - 12k^2) \cdot (42k^2 - 24k)$

$\boxed{\dfrac{dp}{dk} = 7.9\cos(14k^3 - 12k^2) \cdot (42k^2 - 24k)}$

Alternate form: $\dfrac{dp}{dk} = (331.8k^2 - 189.6k)\cos(14k^3 - 12k^2)$

17. a. $p(2) \approx \boxed{10.376 \text{ thousand people}}$

b. $g(10.376) \approx \boxed{19.634 \text{ garbage trucks}}$

c. $\left.\dfrac{dp}{dt}\right|_{t=2} \approx \boxed{0.191 \text{ thousand people per year}}$

d. $\left.\dfrac{dg}{dp}\right|_{p \approx 10.376} \approx \boxed{1.677 \text{ trucks per thousand people}}$

e. $\left.\dfrac{dg}{dt}\right|_{t=2} = \left.\dfrac{dp}{dt} \cdot \dfrac{dg}{dp}\right|_{t=2} \approx \boxed{0.320 \text{ garbage trucks per year}}$

19. a. The functions S and u can be composed to form the required function: $S(u(x))$

$\boxed{S(u(x)) = 0.75\sqrt{-2.3x^2 + 53x + 250} + 1.8 \text{ million dollars gives the sales generated when } x \text{ is the number of months since the beginning of the ad campaign.}}$

b. $\dfrac{dS}{dx} = \dfrac{dS}{du} \cdot \dfrac{du}{dx}$

$= 0.75(0.5u^{-0.5}) \cdot (-4.6x + 53)$

$\boxed{\dfrac{dS}{dx} = 0.75(0.5(-2.3x^2 + 53x + 250)^{-0.5}) \cdot (-4.6x + 53) \text{ million dollars per month}}$

Alternate form: $S'(x) = \dfrac{-1.725x + 19.875}{\sqrt{-2.3x^2 + 53x + 250}}$

c. $\left.\dfrac{dS}{dx}\right|_{x=12} \approx \boxed{-0.035 \text{ million dollars per month}}$

21. a. The functions r and f can be composed to create the required function: $r(f(t))$

$r(f(7)) \approx \boxed{\$1049.32}$

b. $\dfrac{dr}{dt} = \dfrac{dr}{df} \cdot \dfrac{df}{dt}$

$= (-0.0018f^2 - 0.36f) \cdot (0.369t^2 - 6.6t + 22.2)$

$= (-0.0018(f(t))^2 - 0.36f(t)) \cdot (0.369t^2 - 6.6t + 22.2)$

where $f(t) = 0.123t^3 - 3.3t^2 + 22.2t + 55.72$

Alternate form:

$\dfrac{dr}{dt} = (-0.0018(0.123t^3 - 3.3t^2 + 22.2t + 55.72)^2 - 0.36(0.123t^3 - 3.3t^2 + 22.2t + 55.72)) \cdot (0.369t^2 - 6.6t + 22.2)$

$\approx -569.379 - 107.007t + 94.049t^2 - 6.623t^3 - 0.713t^4 + 0.007t^5 + 0.003t^6 + 1.8 \cdot 10^{-4}t^7 - 1.0 \cdot 10^{-5}t^8$

$\left.\dfrac{dr}{dt}\right|_{t=7} \approx \boxed{\$17.83 \text{ per month}}$

23. a. The functions f and x can be composed to create the required function: $f(x(a))$
   $\boxed{f(x(a)) = 0.89 + 0.495 \ln(a+10) \text{ inches gives the average length of the outer ear of a man whose age is } a \text{ years, data from } 0 \le a \le 70.}$

   b. $\dfrac{df}{da} = \dfrac{df}{dx} \cdot \dfrac{dx}{da}$

   $= \dfrac{0.495}{x} \cdot 1$

   $= \dfrac{0.495}{a+10}$

   $\boxed{\dfrac{df}{da} = \dfrac{0.495}{a+10} \text{ inches per year gives the rate of change of the length of the outer ear of a man whose age is } a \text{ years, data from } 0 \le a \le 70.}$

   c. average ear length: $f(x(20)) \approx \boxed{2.574 \text{ inches}}$

   ROC: $\left. \dfrac{df}{da} \right|_{a=20} \approx \boxed{0.017 \text{ inches per year}}$

   $\boxed{\text{The average outer ear length for a 20-year-old male is approximately 2.6 inches and is increasing by approximately 0.02 inches per year.}}$

25. a. $u(x) \approx 7.763x^2 + 47.447x + 1945.893$ units gives production during the xth quarter, based on a four-year period.

   b. $C(u(x)) \approx 196.3 + 44.5 \ln \left| 7.763x^2 + 47.447x + 1945.893 \right|$ dollars gives the production cost for week x, based on a four-year period.

   c. 18th quarter: $C(u(18)) \approx 578.034$ → $\boxed{\$578.03}$

   20th quarter: $C(u(20)) \approx 583.428$ → $\boxed{\$583.43}$

   d. $\dfrac{dC}{dx} = \dfrac{dC}{du} \cdot \dfrac{du}{dx}$

   $\approx \dfrac{44.5}{u} \cdot (15.526x + 47.447)$

   $\approx \dfrac{44.5}{7.763x^2 + 44.447x + 1945.893} \cdot (15.526x + 47.447)$ where $x > 0$

   $\boxed{\dfrac{dC}{dx} = \dfrac{44.5(15.526x + 47.447)}{7.763x^2 + 44.447x + 1945.893} \text{ dollars per quarter}}$

   $\boxed{\text{The cost is always increasing because the rate of change is positive for all positive } x.}$

27. Answers will vary but may be similar to the following:
   Unless the output units of the inside function match the input units of the outside function the composed function does not make sense.
   Suppose the outside function, $C(n)$, models cost as a function of the number of items purchased. A function, $n(t)$ dollars where t is day t of a vacation would not be suitable for composition with $C(n)$. Matching input and output variable letters is not sufficient for composition of functions in a context.

29. a. $m(4.5) \approx 4.583$ hundred dollars

$m(10.5) \approx 18.340$ hundred dollars

Reversing the input and output values gives the two points $(4.583, 4.5)$ and $(18.340, 10.5)$. Because m is an exponential function, its inverse should be a log function. Using the inverted points, the log model is

$x(m) \approx -2.087 + 4.327 \ln m$ thousand dollars gives the amount spent by the consumer for all personal consumption where m hundred dollars is the amount spent by that consumer on his or her motor vehicles.

b. The function n and the inverse function of m can be composed to create the function: $n(x(m))$

$n(x(m)) = -1.1 + 1.64 \ln \left| -2.087 + 4.327 \ln m \right|$ thousand dollars gives the amount spent by a consumer on nondurable goods who also spends m hundred dollars on his or her motor vehicles.

c. spending: $n(x(3.4)) \approx$ 0.812 thousand dollars

$$\frac{dn}{dm} = \frac{dn}{dx} \cdot \frac{dx}{dm}$$

$$\approx \frac{1.64}{x} \cdot \frac{4.327}{m}$$

$$\approx \frac{1.64}{-2.087 + 4.327 \ln m} \cdot \frac{4.327}{m}$$

Alternate form: $\dfrac{dn}{dm} \approx \dfrac{7.096}{-2.087m + 4.327m \ln m}$

ROC: $\left. \dfrac{dn}{dm} \right|_{m=3.4} \approx 0.651$ thousand dollars per hundred dollars (thousand dollars spent on nondurable goods per hundred dollars spent on motor vehicles)

A consumer who spends \$340 on his or her motor vehicles will also spend \$812 on nondurable goods. At the \$340 level of spending on motor vehicles, the amount spent on nondurable goods is increasing by \$651 per hundred dollar increase in spending on motor vehicles.

Section 3.4 Rates of Change of Composite Functions (pages 223–225)

1. inside: $g(x) = 4x^2 + 3$

   outside: $f(g) = 6g^5$

   derivative: $f'(x) = f'(g) \cdot g'(x)$

   $= 30g^4 \cdot 8x$

   $= 30(4x^2 + 3)^4 \cdot 8x$

   $f'(x) = 240x(4x^2 + 3)^4$

3. inside: $g(x) = 5x^2 + 8$

   outside: $f(g) = 2 \ln g$

derivative: $f'(x) = f'(g) \cdot g'(x)$

$$= \frac{2}{g} \cdot 10x$$

$$\boxed{f'(x) = \frac{2}{5x^2+8} \cdot 10x}$$

Alternate form: $f'(x) = \frac{20x}{5x^2+8}$

5. inside: $g(x) = 0.7x$
   outside: $f(g) = 17e^g + \pi$
   derivative: $f'(x) = f'(g) \cdot g'(x)$
   $$= 17e^g \cdot 0.7$$
   $$= 17e^{0.7x} \cdot 0.7$$
   $$\boxed{f'(x) = 11.9e^{0.7x}}$$

7. rewrite $f(x) = \frac{3}{(x^5-1)^3}$ as $f(x) = 3(x^5-1)^{-3}$
   inside: $g(x) = x^5 - 1$
   outside: $f(g) = 3g^{-3}$
   derivative: $f'(x) = f'(g) \cdot g'(x)$
   $$= -9g^{-4}(5x^4)$$
   $$= -9(x^5-1)^{-4} \cdot 5x^4$$
   $$\boxed{f'(x) = -45x^4(x^5-1)^{-4}}$$

   Alternate form: $f'(x) = \frac{-45x^4}{(x^5-1)^4}$

9. rewrite $f(x) = 3\sqrt{x^3 + 2\ln x}$ as $f(x) = 3(x^3 + 2\ln x)^{\frac{1}{2}}$
   inside: $g(x) = x^3 + 2\ln x$
   outside: $f(g) = 3g^{\frac{1}{2}}$
   derivative: $f'(x) = f'(g) \cdot g'(x)$
   $$= 1.5g^{-\frac{1}{2}}\left(3x^2 + \frac{2}{x}\right)$$
   $$\boxed{f'(x) = 1.5(x^3 + 2\ln x)^{-\frac{1}{2}}\left(3x^2 + \frac{2}{x}\right)}$$

Alternate form: $f'(x) = \dfrac{1.5\left(3x^2 + \dfrac{2}{x}\right)}{\sqrt{x^3 + 2\ln x}}$

$= \dfrac{1.5\left(3x^2 + \dfrac{2}{x}\right)\sqrt{x^3 + 2\ln x}}{x^3 + 2\ln x}$

11. inside: $g(x) = 3.2x + 5.7$
    outside: $f(g) = g^5$
    derivative: $f'(x) = f'(g) \cdot g'(x)$
    $= 5g^4 \cdot 3.2$
    $\boxed{f'(x) = 16(3.2x + 5.7)^4}$

13. rewrite $f(x) = \dfrac{8}{(x-1)^3}$ as $f(x) = 8(x-1)^{-3}$
    inside: $g(x) = x - 1$
    outside: $f(g) = 8g^{-3}$
    derivative: $f'(x) = f'(g) \cdot g'(x)$
    $= -24g^{-4} \cdot 1$
    $\boxed{f'(x) = -24(x-1)^{-4}}$
    Alternate form: $f'(x) = \dfrac{-24}{(x-1)^4}$

15. rewrite $f(x) = \sqrt{(x^2 - 3x)}$ as $f(x) = (x^2 - 3x)^{\frac{1}{2}}$
    inside: $g(x) = x^2 - 3x$
    outside: $f(g) = g^{\frac{1}{2}}$
    derivative: $f'(x) = f'(g) \cdot g'(x)$
    $= \dfrac{1}{2}g^{-\frac{1}{2}}(2x - 3)$
    $\boxed{f'(x) = \dfrac{1}{2}(x^2 - 3x)^{-\frac{1}{2}}(2x - 3)}$
    Alternate form: $f'(x) = \dfrac{2x - 3}{2(x^2 - 3x)^{\frac{1}{2}}}$
    $= \dfrac{2x - 3}{2\sqrt{x^2 - 3x}}$
    $= \dfrac{(2x - 3)\sqrt{x^2 - 3x}}{2(x^2 - 3x)}$

17. inside: $g(x) = 35x$
    outside: $f(g) = \ln g$
    derivative: $f'(x) = f'(g) \cdot g'(x)$
    $$= \frac{1}{g} \cdot 35$$
    $$= \frac{1}{35x} \cdot 35$$
    $$\boxed{f'(x) = \frac{1}{x}}$$
    Alternate form: $f'(x) = x^{-1}$

19. inside: $g(x) = 16x^2 + 37x$
    outside: $f(g) = \ln g$
    derivative: $f'(x) = f'(g) \cdot g'(x)$
    $$= \frac{1}{g}(32x + 37)$$
    $$\boxed{f'(x) = \frac{32x + 37}{16x^2 + 37x}}$$

21. inside: $g(x) = 0.6x$
    outside: $f(g) = 72e^g$
    derivative: $f'(x) = f'(g) \cdot g'(x)$
    $$= 72e^g \cdot 0.6 \quad \text{or} \quad 72(0.6e^g)$$
    $$\boxed{f'(x) = 43.2e^{0.6x}}$$

23. inside: $g(x) = 0.08x$
    outside: $f(g) = 1 + 58e^g$
    derivative: $f'(x) = f'(g) \cdot g'(x)$
    $$= 58e^g \cdot 0.08 \quad \text{or} \quad 58(0.08e^g)$$
    $$\boxed{f'(x) = 4.64e^{0.08x}}$$

25. rewrite $f(x) = \frac{12}{1 + 18e^{0.6x}}$ as $f(x) = 12(1 + 18e^{0.6x})^{-1}$
    inside: $h(x) = 0.6x$
    intermediate: $g(h) = 1 + 18e^h$
    outside: $f(g) = 12g^{-1}$
    derivative: $f'(x) = f'(g) \cdot g'(h) \cdot h'(x)$
    $$= -12g^{-2} \cdot 18e^h \cdot 0.6$$
    $$= -12(1 + e^{0.6x})^{-2} \cdot 18e^{0.6x} \cdot 0.6$$

$$\boxed{f'(x) = -129.6e^{0.6}(1+18e^{0.6x})^{-2}}$$

Alternate form: $f'(x) = \dfrac{-129.6e^{0.6}}{(1+18e^{0.6x})^2}$

27. inside: $g(x) = \ln x$
    outside: $f(g) = 2^g$
    derivative: $f'(x) = f'(g) \cdot g'(x)$
    $$= (\ln 2 \cdot 2^g) \cdot \dfrac{1}{x}$$
    $$\boxed{f'(x) = (\ln 2 \cdot 2^{\ln x}) \cdot \dfrac{1}{x}}$$
    Alternate form: $f'(x) = \dfrac{\ln 2 \cdot 2^{\ln x}}{x}$

29. inside: $g(x) = 2x + 5$
    outside: $f(g) = 3\sin g + 7$
    derivative: $f'(x) = f'(g) \cdot g'(x)$
    $$= 3\cos g \cdot 2$$
    $$\boxed{f'(x) = 6\cos(2x+5)}$$

31. $f(x) = 4\sin(5\ln x + 7) + 5x$ can be rewritten as $f(x) = k(x) + j(x)$ where
    $j(x) = 5x$ and
    $k(x) = 4\sin(5\ln x + 7)$ which can be written as the composition of
    inside: $g(x) = 5\ln x + 7$
    outside: $k(g) = 4\sin g$
    derivative: $f'(x) = k'(g) \cdot g'(x) + j'(x)$
    $$= 4\cos g \cdot \dfrac{5}{x} + 5$$
    $$\boxed{f'(x) = 4\cos(5\ln x + 7) \cdot \dfrac{5}{x} + 5}$$
    Alternate form: $f'(x) = \dfrac{20\cos(5\ln x + 7)}{x} + 5$

33. a. $p(t) = 2.111e^{0.04t}$ can be written as the composition of
    inside: $g(t) = 0.04t$
    outside: $p(g) = 2.111e^g$
    $p'(t) = p'(g) \cdot g'(t)$
    $$= 2.111e^g \cdot 0.04$$
    $$= 2.111e^{0.04t} \cdot 0.04$$
    $$\boxed{p'(t) \approx 0.084e^{0.04t}}$$

b. $p'(30) \approx \boxed{0.280 \text{ million children per year}}$

35. a. $f(x) = 2.5\ln(113.17x^{1.222}) + 33.3$ can be written as the composition of
inside: $g(x) = 113.17x^{1.222}$
outside: $f(g) = 2.5\ln g + 33.3$
$f'(x) = f'(g) \cdot g'(x)$
$= 2.5 \dfrac{1}{g} \cdot 113.17(1.222x^{0.222})$
$= 2.5 \dfrac{1}{113.17x^{1.222}} \cdot 113.17(1.222x^{0.222})$

$\boxed{f'(x) \approx \dfrac{345.734x^{0.222}}{113.17x^{1.222}} \text{ percentage points per billion passengers gives the rate of change in the percentage of total revenue generated by enplaned passengers where } x \text{ billion is the number of enplaned passengers.}}$

Alternate form: $f'(x) = 3.055x^{0.222-1.222}$
$= 3.055x^{-1}$
$= \dfrac{3.055}{x}$

b. $f(0.5) \approx \boxed{43.005\%}$

c. $f'(0.5) = \boxed{6.11 \text{ percentage points per billion passengers}}$

37. a. $s(0.75) \approx \boxed{173.815 \text{ thousand gallons}}$

b. $s(x) = 597.3(0.921^{4x+12})$ can be written as the composition of
inside: $g(x) = 4x + 12$
outside: $s(g) = 597.3(0.921^g)$
$s'(x) = s'(g) \cdot g'(x)$
$= 597.3(\ln 0.921 \cdot 0.921^g) \cdot 4$
$= 597.3(\ln 0.921 \cdot 0.921^{4x+12}) \cdot 4$
$\approx -196.620(0.921^{4x+12})$

$s'(0.75) \approx \boxed{-57.217 \text{ thousand gallons per dollar}}$

39. a. $f(x) = 37\sin(0.0172x - 1.737) + 25$ can be written as the composition of
inside: $g(x) = 0.0172x - 1.737$
outside: $f(g) = 37\sin g + 25$
$f'(x) = f'(g) \cdot g'(x)$
$= 37\cos g \cdot 0.0172$
$= 37\cos(0.0172x - 1.737) \cdot 0.0172$
$\boxed{f'(x) = 0.6364\cos(0.0172x - 1.737)\text{°F per day}}$

b. $f'(180) \approx \boxed{0.134\text{°F per day}}$

> Near July 1 (180 days into the calendar year), the normal mean temperature in Fairbanks, Alaska is increasing by approximately 0.13°F per day.

41. a. $s(x) \approx \dfrac{1342.077}{1+36.797e^{-0.259x}}$ calls gives the total number of calls received at a sheriff's office during a 24-hour period, where x is the number of hours since 5 A.M., data from 8 A.M. to 5 A.M.

   b. $s(x) \approx \dfrac{1342.077}{1+36.797e^{-0.259x}}$ can be rewritten as $s(x) \approx 1342.077(1+36.797e^{-0.259x})^{-1}$ which can be written as the composition of

   inside: $h(x) \approx -0.259x$

   intermediate: $g(h) \approx 1+36.797e^{h}$

   outside: $s(g) \approx 1342.077 g^{-1}$

   $s'(x) = s'(g) \cdot g'(h) \cdot h'(x)$

   $\approx -1342.077 g^{-2} \cdot 36.797 e^{h} \cdot (-0.259)$

   $= -1342.077(1+36.797e^{-0.259x})^{-2} \cdot 36.797 e^{-0.259x} \cdot (-0.259)$

   $\approx 12{,}783.365 e^{-0.259x}(1+36.797e^{-0.259x})^{-2}$

   Alternate form: $s'(x) \approx \dfrac{12{,}783.365 e^{-0.259x}}{(1+36.797e^{-0.259x})^{2}}$

   noon: $s'(7) \approx$ 42.489 calls per hour

   10 P.M.: $s'(17) \approx$ 74.450 calls per hour

   midnight: $s'(19) \approx$ 58.038 calls per hour

   4 A.M.: $s'(23) \approx$ 27.653 calls per hour

   c. Answers will vary but might include the following:
   Because the function s gives total number of calls received in the x hours since 5 A.M., the derivative of s gives the number of calls being received in the hour centered at time x. These rates of change might be useful in letting schedulers know the appropriate number of dispatchers and responders needed to handle the volume of calls at time x.

Section 3.5 Rates of Change for Functions that can be Multiplied (pages 232–234)

1. $h'(2) = f'(2) \cdot g(2) + f(2) \cdot g'(2)$

   $= -1.5 \cdot 4 + 6 \cdot 3$

   $= \boxed{12}$

3. a. i. In 2012, there are 75,000 households in the city.
   ii. In 2012, the number of households in the city is decreasing by 1200 per year.
   iii. In 2012, 90% of the households in the city have multiple computers.
   iv. In 2012, the proportion of households in the city with multiple computers is increasing by 5% per year.

3.5 Rates of Change for Functions that can be Multiplied

b. input: t is the number of years since 2010
output: $n(t)$ is the number of households with computers

c. $n(2) = h(2) \cdot c(2)$
$= $ 67,500 households

In 2012, 67,500 households in the city have multiple computers.

$n'(2) = h'(2) \cdot c(2) + h(2) \cdot c'(2)$
$= -1200 \cdot 0.9 + 75000 \cdot 0.05$
$= $ 2670 households per year

In 2012, the number of households in the city with multiple computers is increasing by 2,670 households per year.

5. rewrite $s(x) = 15 + \dfrac{2.6}{x+1}$ as $s(x) = 15 + 2.6(x+1)^{-1}$

$s'(x) = 0 - 2.6(x+1)^{-2} \cdot 1$ by the difference and chain rules
$= -2.6(x+1)^{-2}$

Alternate form: $s'(x) = \dfrac{-2.6}{(x+1)^2}$

$n(x) = 100 + 0.25x^2$
$n'(x) = 0.5x$

a. i. $s(10) \approx 15.236 \;\rightarrow\;$ $15.24 (or dollars/share)

$s'(10) \approx -0.021 \;\rightarrow\;$ −$0.02 per week (or dollars/share per week)

(The answer key in the text erroneously states "per share" instead of "per week".)

Ten weeks after the first offering of a company's stock, the value of one share is $15.24 and is decreasing by $0.02 per week.

ii. $n(10) = $ 125 shares
$n'(10) = $ 5 shares per week

Ten weeks after the first offering of a company's stock, an investor owns 125 shares of stock and is increasing his shares by 5 shares per week.

iii. (The answer key in the text used rounded numbers in the calculation.)
$v(10) = s(10) \cdot n(10)$
$\approx 1904.545 \;\rightarrow\;$ $1904.55

$v'(10) = s'(10) \cdot n(10) + s(10) \cdot n'(10)$
$\approx 73.496 \;\rightarrow\;$ $73.50 per week

Ten weeks after the first offering of a company's stock, the value of an investor's holding is approximately $2170.50 and is increasing by approximately $59.96 per week.

b. $v'(x) = s'(x) \cdot n(x) + s(x) \cdot n'(x)$

$$\boxed{v'(x) = -26(x+1)^{-2} \cdot (100 + 0.25x^2) + (15 + 26(x+1)^{-1}) \cdot 0.5x \text{ dollars per week}}$$

Alternate form: $v'(x) = \dfrac{-26(100 + 0.25x^2)}{(x+1)^2} + \left(15 + \dfrac{26}{x+1}\right) \cdot 0.5x$

$= \dfrac{-2600 - 6.5x^2}{(x+1)^2} + \dfrac{13x}{x+1} + 7.5x$

$= \dfrac{7.5x^3 + 21.5x^2 + 20.5x - 2600}{(x+1)^2}$

7. Total yield can be expressed as
   total yield = (acres of corn)·(yield per acre)

   Rate of change of total yield can be calculated as

   $\begin{pmatrix} \text{ROC of} \\ \text{total yield} \end{pmatrix} = \begin{pmatrix} \text{ROC of} \\ \text{acres of corn} \end{pmatrix} \cdot (\text{yield per acre}) + (\text{acres of corn}) \cdot \begin{pmatrix} \text{ROC of} \\ \text{yield per acre} \end{pmatrix}$

   $= (50 \text{ acres per year}) \cdot (130 \text{ bushels/acre}) + (500 \text{ acres}) \cdot (5 \text{ bushels/acre per year})$

   $= \boxed{9000 \text{ bushels per year}}$

   Alternate set-up:
   Let $a(x)$ acres represent corn acreage in year x.
   Let $y(x)$ bushels/acre represent corn yield in year x.
   Total corn yield in year x can be expressed as
   $c(x) = a(x) \cdot y(x)$
   Rate of change of total corn yield in year x can be calculated as
   $c'(x) = a'(x) \cdot y(x) + a(x) \cdot y'(x)$
   $= 50 \cdot 130 + 500 \cdot 5$
   $= 9000$ bushels per year

9. a. Registered voters who would vote today can be calculated as
      (registered voters)·(proportion voting)
      $= (17{,}000 \text{ voters}) \cdot (0.48)$
      $= \boxed{8160 \text{ voters}}$

   b. Voters who would vote for candidate A today can be calculated as
      (today's voters)·(proportion voting for candidate A)
      $= (8160 \text{ voters}) \cdot (0.57)$
      $= 4651.2 \quad \rightarrow \quad \approx \boxed{4651 \text{ voters}}$

   c. Rate of change of votes for candidate A can be calculated as

      $\begin{pmatrix} \text{ROC of votes} \\ \text{for candidate A} \end{pmatrix}$

      $= \begin{pmatrix} \text{ROC of} \\ \text{voters} \end{pmatrix} \cdot (\text{proportion for A}) + (\text{voters}) \cdot \begin{pmatrix} \text{ROC of} \\ \text{proportion for A} \end{pmatrix}$

      $= (17{,}000 \text{ voters} \cdot 0.07 \text{ per week}) \cdot (0.57) + (17{,}000 \text{ voters} \cdot 0.48) \cdot (-0.03 \text{ per week})$

      $= 433.5 \quad \rightarrow \quad \approx \boxed{434 \text{ votes per week}}$

11. a. $f(x) = (5x^2 - 3) \cdot 1.2^x$
    b. $f'(x) = g'(x) \cdot h(x) + g(x) \cdot h'(x)$
    $$\boxed{f'(x) = 10x \cdot 1.2^x + (5x^2 - 3) \cdot (\ln 2 \cdot 1.2^x)}$$

13. a. $f(x) = (4x^2 - 25) \cdot (20 - 7\ln x)$
    b. $f'(x) = g'(x) \cdot h(x) + g(x) \cdot h'(x)$
    $$\boxed{f'(x) = 8x \cdot (20 - 7\ln x) + (4x^2 - 25) \cdot \frac{-7}{x}}$$

    Alternate form: $f'(x) = 8x \cdot (20 - 7\ln x) - \frac{28x^2 - 175}{x}$
    $$= 160x - 56x \ln x - \frac{28x^2 - 175}{x}$$
    $$= \frac{132x^2 - 56x^2 \ln x + 175}{x}$$

15. a. $f(x) = 2e^{1.5x} \cdot 2(1.5^x)$
    $$\boxed{f(x) = 4e^{1.5x} 1.5^x}$$
    b. $f'(x) = g'(x) \cdot h(x) + g(x) \cdot h'(x)$
    $$\boxed{f'(x) = (4e^{1.5x} \cdot 1.5) \cdot (1.5^x) + (4e^{1.5x}) \cdot (\ln 1.5 \cdot 1.5^x)}$$
    Alternate form: $f'(x) = 6e^{1.5x} 1.5^x + 4\ln 1.5 \cdot e^{1.5x} 1.5^x$

17. a. $f(x) = \left(6e^{-x} + \ln x\right) \cdot 4x^{2.1}$
    b. $f'(x) = g'(x) \cdot h(x) + g(x) \cdot h'(x)$
    $$\boxed{f'(x) = \left(-6e^{-x} + \frac{1}{x}\right) \cdot 4x^{2.1} + (6e^{-x} + \ln x) \cdot 8.4x^{1.1}}$$

    Alternate form: $f'(x) = -24x^{2.1}e^{-x} + \frac{4x^{2.1}}{x} + 50.4x^{1.1}e^{-x} + 8.4x^{1.1} \ln x$
    $$= -24x^{2.1}e^{-x} + 4x^{1.1} + 50.4x^{1.1}e^{-x} + 8.4x^{1.1} \ln x$$

19. a. $f(x) = (-3x^2 + 4x - 5) \cdot (0.5x^{-2} - 2x^{0.5})$
    b. $f'(x) = g'(x) \cdot h(x) + g(x) \cdot h'(x)$
    $$\boxed{f'(x) = (-6x + 4) \cdot (0.5x^{-2} - 2x^{0.5}) + (-3x^2 + 4x - 5) \cdot (-x^{-3} - x^{-0.5})}$$

21. (The second paragraph in the activity should read "Suppose that the yearly number of women giving birth at a small hospital in southern Arizona can be modeled as…")
    a. In order to use the percentage function p to convert the number of women giving birth to the number of women using regional analgesia, p must be converted from percentage to proportion by diving by 100.

The number of women using regional analgesia while giving birth can be expressed as

$$f(x) = b(x) \cdot \frac{p(x)}{100}$$

$$\boxed{f(x) = (-0.026x^2 - 3.842x + 538.868) \cdot \frac{0.73(1.2912^x) + 8}{100} \text{ women}}$$

ROC:

$$f'(x) = b'(x) \cdot \frac{p(x)}{100} + b(x) \cdot \frac{p'(x)}{100}$$

$$\boxed{f'(x) = (-0.052x - 3.842) \cdot \frac{0.73(1.2912^x) + 8}{100} + (-0.026x^2 - 3.842x + 538.868) \cdot \frac{0.73 \ln 1.2912(1.2912^x)}{100}}$$

$\boxed{\text{women per year}}$

b. $p'(17) \approx 14.379$ percentage points per year

The rate of change of percentage in 1997 is positive indicating that the percentage is $\boxed{\text{increasing}}$.

c. $b'(17) \approx -4.726$ women

The rate of change of the number of women giving births in 1997 is negative indicating that the number of women giving birth at the Arizona hospital is $\boxed{\text{decreasing}}$.

d. $f'(17) = b'(17) \cdot \frac{p(17)}{100} + b(17) \cdot \frac{p'(17)}{100}$

≈ 63.974 women per year

The rate of change of the number of women who received regional analgesia while giving birth is positive indicating that this number was $\boxed{\text{increasing}}$.

23. a. The population of the Midwest can be expressed as

$$f(x) = p(x) \cdot \frac{m(x)}{100}$$

$$\boxed{f(x) = 203.12e^{0.011x} \cdot \frac{0.002x^2 - 0.213x + 27.84}{100} \text{ million people}}$$

b. $f'(x) = p'(x) \cdot \frac{m(x)}{100} + p(x) \cdot \frac{m'(x)}{100}$

$$\boxed{f'(x) = 203.12(0.011e^{0.011x}) \cdot \frac{0.002x^2 - 0.213x + 27.84}{100} + 203.12e^{0.011x} \cdot \frac{0.004x - 0.213}{100}}$$

$\boxed{\text{million people per decade}}$

c. 2000: $f'(3) \approx \boxed{0.207 \text{ million people per decade}}$

2010: $f'(4) \approx \boxed{0.213 \text{ million people per decade}}$

25. Answers will vary but should include the following points:
Unless the input units of the functions to be multiplied are the same and their alignments match, there is no meaning that can be given to the function that results from multiplication. Even if the output units seem sensible, there is no way to describe the input variable.

Section 3.6 Rates of Change of Product Functions (pages 237–239)

1. $f(x) = (\ln x)e^x$

   $f'(x) = \dfrac{1}{x}e^x + (\ln x)e^x$

   $\boxed{f'(x) = \dfrac{e^x}{x} + (\ln x)e^x}$

3. $f(x) = (3x^2 + 15x + 7)(32x^3 + 49)$

   $\boxed{f'(x) = (6x + 15)(32x^3 + 49) + (3x^2 + 15x + 7)(96x^2)}$

5. $f(x) = (12.8x^2 + 3.7x + 1.2)[29(1.7^x)]$

   $\boxed{f'(x) = (25.6x + 3.7)(29(1.7^x)) + (12.8x^2 + 3.7x + 1.2)(29\ln 1.7(1.7^x))}$

7. $f(x) = (5.7x^2 + 3.5x + 2.9)^3(3.8x^2 + 5.2x + 7)^{-2}$

   $\boxed{\begin{aligned} f'(x) &= 3(5.7x^2 + 3.5x + 2.9)^2(11.4x + 3.5)(3.8x^2 + 5.2x + 7)^{-2} \\ &\quad - 2(5.7x^2 + 3.5x + 2.9)^3(3.8x^2 + 5.2x + 7)^{-3}(7.6x + 5.2) \end{aligned}}$

   Alternate form:

   $f'(x) = \dfrac{3(5.7x^2 + 3.5x + 2.9)^2(11.4x + 3.5)}{(3.8x^2 + 5.2x + 7)^2} - \dfrac{2(5.7x^2 + 3.5x + 2.9)^3(7.6x + 5.2)}{(3.8x^2 + 5.2x + 7)^3}$

9. rewrite $f(x) = \dfrac{12.6(4.8^x)}{x^2}$ as $f(x) = 12.6(4.8^x)x^{-2}$

   $\boxed{f'(x) = 12.6(\ln 4.8)(4.8^x)x^{-2} - 25.2(4.8^x)x^{-3}}$

   Alternate form: $f'(x) = \dfrac{12.6(\ln 4.8)(4.8^x)}{x^2} - \dfrac{25.2(4.8^x)}{x^3}$

11. $f(x) = (79x)\left(\dfrac{198}{1 + 7.68e^{-0.85x}} + 15\right)$

    $\boxed{f'(x) = 79\left(\dfrac{198}{1 + 7.68e^{-0.85x}} + 15\right) - (79x)(198)(1 + 7.68e^{-0.85x})^{-2}(7.68e^{-0.85x})(-0.85)}$

    Alternate form: $f'(x) = 79\left(\dfrac{198}{1 + 7.68e^{-0.85x}} + 15\right) + 102{,}110.976x(1 + 7.68e^{-0.85x})^{-2}e^{-0.85x}$

    $= 79\left(\dfrac{198}{1 + 7.68e^{-0.85x}} + 15\right) + \dfrac{102{,}110.976xe^{-0.85x}}{(1 + 7.68e^{-0.85x})^2}$

13. rewrite $f(x) = \dfrac{430(0.62^x)}{6.42 + 3.3(1.46^x)}$ as $f(x) = 430(0.62^x)\big(6.42 + 3.3(1.46^x)\big)^{-1}$

$$f'(x) = 430(\ln 0.62)(0.62^x)\big(6.42 + 3.3(1.46^x)\big)^{-1} - 1419(0.62^x)\big(6.42 + 3.3(1.46^x)\big)^{-2}(\ln 1.46)(1.46^x)$$

Alternate form: $f'(x) = \dfrac{430(\ln 0.62)(0.62^x)}{6.42 + 3.3(1.46^x)} - \dfrac{1419(0.62^x)(\ln 1.46)(1.46^x)}{(6.42 + 3.3(1.46^x))^2}$

15. rewrite $f(x) = 4x\sqrt{3x+2} + 93$ as $f(x) = 4x(3x+2)^{0.5} + 93$

$$f'(x) = 4(3x+2)^{0.5} + 6x(3x+2)^{-0.5}$$

Alternate form: $f'(x) = 4(3x+2)^{0.5} + 6x(3x+2)^{-0.5}$

$$= 4\sqrt{3x+2} + \dfrac{6x\sqrt{3x+2}}{3x+2}$$

17. rewrite $f(x) = \dfrac{14x}{1 + 12.6e^{-0.73x}}$ as $f(x) = 14x\big(1 + 12.6e^{-0.73x}\big)^{-1}$

$$f'(x) = 14\big(1 + 12.6e^{-0.73x}\big)^{-1} + 128.772x\big(1 + 12.6e^{-0.73x}\big)^{-2} e^{-0.73x}$$

Alternate form: $f'(x) = \dfrac{14}{1 + 12.6e^{-0.73x}} + \dfrac{128.772xe^{-0.73x}}{(1 + 12.6e^{-0.73x})^2}$

19. a. The rate of change of profit can be expressed as
$$P'(q) = 72e^{-0.2q} + 72q(-0.2e^{-0.2q}) \text{ dollars per unit}$$
Alternate form: $P'(q) = 72e^{-0.2q} - 14.4qe^{-0.2q}$

   b. Solving $P'(q) = 0$ yields $q = \boxed{5 \text{ units}}$

   c. $P(5) \approx 132.437 \;\to\; \boxed{\$132.44}$

21. a. Average profit is the profit for one unit:
$$\dfrac{\text{thousand dollars}}{\text{million units}} = \dfrac{1000 \text{ dollars}}{1{,}000{,}000 \text{ units}} = 0.001 \text{ dollars per unit}.$$
Average profit can be expressed as
$$\overline{P}(q) = 0.001 \dfrac{P(q)}{q}$$

$$\overline{P}(q) = 0.001 \dfrac{30 + 60\ln q}{q} \text{ dollars}$$

   b. $P(10) \approx \boxed{\$168.155 \text{ thousand}}$
   $\overline{P}(10) \approx \boxed{\$0.017}$

c. $P'(q) = \dfrac{60}{q}$

$\overline{P}'(q) = 0.001\left[\dfrac{60}{q}\cdot\dfrac{1}{q} + (30 + 60\ln q)\left(-\dfrac{1}{q^2}\right)\right]$

$P'(10) =$ $\boxed{\$6 \text{ thousand per million units}}$

$\overline{P}'(10) \approx$ $\boxed{-\$0.001 \text{ per million units}}$

d. Answers will vary:
If a manager is interested in maximizing average profit, he should pay attention to the rate of change of average profit. Maximum average profit generally occurs at a lower production level than maximum profit.

23. a. $R(x) = 6250(0.929^x)x$ dollars gives the revenue from the sale of Blu-ray movies at an average price of x dollars.

b. $P(x) = 6250(0.929^x)(x-10)$ dollars gives the profit from the sale of Blu-ray movies at an average price of x dollars.

c. $R'(x) = 6250(\ln 0.929)(0.929^x)x + 6250(0.929^x)$
$P'(x) = 6250(\ln 0.929)(0.929^x)(x-10) + 6250(0.929^x)$

Price	Rate of change of revenue (dollars per dollar)	Rate of change of profit (dollars per dollar)
$13	102.198	1869.198
$14	-69.212	1572.331
$20	-677.629	377.600
$21	-727.548	252.760
$22	-766.962	143.743

d. Answers will vary:
There is a range of prices beginning near $14 for which the rate of change of revenue is negative (revenue is decreasing) while the rate of change of profit is positive (profit is increasing).

25. a. $p(x) \approx 430.073(1.300^x)$ dollars gives the average income for a painting job where x is the number of years since 2004, based on years between 2004 and 2010.

b. $t(x) = j(x) \cdot p(x)$

$\boxed{t(x) \approx \dfrac{104.25}{x} \cdot 430.073(1.300^x) \text{ dollars gives the painter's yearly income where } x \text{ is the number of years since 2004.}}$

Alternate form: $t(x) \approx 44{,}835.085 x^{-1}(1.300^x)$

c. $\boxed{t'(x) \approx -44{,}835.085 x^{-2}(1.300^x) + 44{,}835.085 x^{-1}(\ln 1.300)(1.300^x) \text{ dollars per year}}$

Alternate form: $t'(x) \approx \dfrac{-44{,}835.085(1.300^x)}{x^2} + \dfrac{44{,}835.085(\ln 1.300)(1.300^x)}{x}$

d. income: $t(6) \approx$ $\boxed{\$36{,}062.80}$

ROC: $t'(6) \approx$ $\boxed{\$3{,}450.19 \text{ per year}}$

Section 3.7 Limits of Quotients and L'Hôpital's Rule (pages 244–245)

1. $\lim\limits_{x \to 2}(2x^3 - 3^x) = 2 \cdot 2^3 - 3^2$ by replacement and limit rules

    $= \boxed{7}$

3. $\lim\limits_{x \to 0}\left[e^x - \ln(x+1)\right] = e^0 - \ln(0+1)$ by replacement and limit rules

    $= \boxed{1}$

5. $\lim\limits_{x \to 0^+} \dfrac{1}{\ln x} = \dfrac{1}{\lim\limits_{x \to 0^+}(\ln x)}$ by the Quotient Rule

    using end-behavior analysis:

    as $x \to 0^+$, $\ln x \to -\infty$

    as $\ln x \to -\infty$, $\dfrac{1}{\ln x} \to 0$

    $\lim\limits_{x \to 0^+} \dfrac{1}{\ln x} = \boxed{0}$

7. $\lim\limits_{n \to 1} \dfrac{\ln n}{n-1} = \dfrac{\ln 1}{1-1}$ has the indeterminate form $\boxed{\dfrac{0}{0}}$

    $\lim\limits_{n \to 1} \dfrac{\ln n}{n-1} = \lim\limits_{n \to 1} \dfrac{\frac{1}{n}}{1}$ by L'Hôpital's Rule

    $= \lim\limits_{n \to 1} \dfrac{1}{n}$ by rewritting

    $= \boxed{1}$

9. $\lim\limits_{x \to 1} \dfrac{x^4 - 1}{x^3 - 1} = \dfrac{1^4 - 1}{1^3 - 1}$ has the indeterminate form $\boxed{\dfrac{0}{0}}$

    $\lim\limits_{x \to 1} \dfrac{x^4 - 1}{x^3 - 1} = \lim\limits_{x \to 1} \dfrac{4x^3}{3x^2}$ by L'Hôpital's Rule

    $= \boxed{\dfrac{4}{3}}$

11. $\lim\limits_{x \to 2} \dfrac{3x - 6}{x + 2} = \dfrac{0}{4}$ by replacement and limit rules

    $= \boxed{0}$

13. $\lim\limits_{x\to 5}\dfrac{(x-1)^{0.5}-2}{x^2-25} = \dfrac{(5-1)^{0.5}-2}{5^2-25}$ has the indeterminate form $\boxed{\dfrac{0}{0}}$

$\lim\limits_{x\to 5}\dfrac{(x-1)^{0.5}-2}{x^2-25} = \lim\limits_{x\to 5}\dfrac{0.5(x-1)^{-0.5}}{2x}$ by L'Hôpital's Rule

$\qquad\qquad\qquad\qquad = \boxed{0.025}$

15. $\lim\limits_{x\to 2}\dfrac{2x^2-5x+2}{5x^2-7x-6} = \dfrac{2\cdot 2^2-5\cdot 2+2}{5\cdot 2^2-7\cdot 2-6}$ has the indeterminate form $\boxed{\dfrac{0}{0}}$

$\lim\limits_{x\to 2}\dfrac{2x^2-5x+2}{5x^2-7x-6} = \lim\limits_{x\to 2}\dfrac{4x-5}{10x-7}$ by L'Hôpital's Rule

$\qquad\qquad\qquad\qquad = \boxed{\dfrac{3}{13}}$

17. $\lim\limits_{x\to\infty}\dfrac{3x^2+2x+4}{5x^2+x+1}$ has the indeterminate form $\dfrac{\infty}{\infty}$

$\lim\limits_{x\to\infty}\dfrac{3x^2+2x+4}{5x^2+x+1} = \lim\limits_{x\to\infty}\dfrac{6x+2}{10x+1}$ by L'Hôpital's Rule

$\lim\limits_{x\to\infty}\dfrac{6x+2}{10x+1}$ has the indeterminate form $\dfrac{\infty}{\infty}$

$\lim\limits_{x\to\infty}\dfrac{6x+2}{10x+1} = \lim\limits_{x\to\infty}\dfrac{6}{10}$ by L'Hôpital's Rule

$\qquad\qquad\qquad = \dfrac{6}{10}$

$\qquad\qquad\qquad = \boxed{\dfrac{3}{5}}$

19. $\lim\limits_{x\to\infty}\dfrac{3x^4}{5x^3+6}$ has the indeterminate form $\boxed{\dfrac{\infty}{\infty}}$

$\lim\limits_{x\to\infty}\dfrac{3x^4}{5x^3+6} = \lim\limits_{x\to\infty}\dfrac{12x^3}{15x^2}$ by L'Hôpital's Rule

$\qquad\qquad\qquad = \lim\limits_{x\to\infty}\dfrac{12x}{15}$ by cancellation

$\qquad\qquad\qquad = \boxed{\infty}$

21. $\lim\limits_{t\to 0^+}(\sqrt{t}\cdot\ln t)$ has the indeterminate form $\boxed{0\cdot\infty}$

rewrite as $\lim\limits_{t\to 0^+}\dfrac{\ln t}{t^{-0.5}}$ which has the indeterminate form $\dfrac{\infty}{\infty}$

$$\lim_{t\to 0^+}(\sqrt{t}\cdot \ln t) = \lim_{t\to 0^+}\frac{\ln t}{t^{-0.5}} \quad \text{by rewritting}$$

$$= \lim_{t\to 0^+}\frac{\frac{1}{t}}{-0.5t^{-1.5}} \quad \text{by L'Hôpital's Rule}$$

$$= \lim_{t\to 0^+}\frac{t^{-1}}{-0.5t^{-1.5}} \quad \text{by rewritting}$$

$$= \lim_{t\to 0^+}\frac{t^{0.5}}{-0.5} \quad \text{by rewritting}$$

$$= \boxed{0}$$

23. $\lim_{x\to\infty} x^2 e^{-x}$ has the indeterminate form $\boxed{\infty\cdot 0}$

rewrite as $\lim_{x\to\infty}\frac{x^2}{e^x}$ which has the indeterminate form $\frac{\infty}{\infty}$

$$\lim_{x\to\infty} x^2 e^{-x} = \lim_{x\to\infty}\frac{x^2}{e^x} \quad \text{by rewritting}$$

$$= \lim_{x\to\infty}\frac{2x}{e^x} \quad \text{by L'Hôpital's Rule}$$

$\lim_{x\to\infty}\frac{2x}{e^x}$ has the indeterminate form $\frac{\infty}{\infty}$

$$\lim_{x\to\infty}\frac{2x}{e^x} = \lim_{x\to\infty}\frac{2}{e^x} \quad \text{by L'Hôpital's Rule}$$

$$= \boxed{0}$$

25. $\lim_{x\to 0}(3x)\left(\frac{2}{e^x}\right) = (3\cdot 0)\cdot\frac{2}{e^0} \quad \text{by replacement and limit rules}$

$$= 0\cdot 2$$

$$= \boxed{0}$$

27. $\lim_{x\to\infty} 3(0.6^x)(\ln x)$ has the indeterminate form $\boxed{0\cdot\infty}$

rewrite as $\lim_{x\to\infty}\frac{3\ln x}{0.6^{-x}}$ which has the indeterminate form $\frac{\infty}{\infty}$

$$\lim_{x\to\infty} 3(0.6^x)(\ln x) = \lim_{x\to\infty}\frac{3\ln x}{0.6^{-x}} \quad \text{by rewritting}$$

$$= \lim_{x\to\infty}\frac{\frac{3}{x}}{-(\ln 0.6)0.6^{-x}} \quad \text{by L'Hôpital's Rule}$$

$$= \lim_{x\to\infty}\frac{3}{-(\ln 0.6)x 0.6^{-x}} \quad \text{by rewritting}$$

as $x \to \infty$, $0.6^{-x} \to \infty$ so $x0.6^{-x} \to \infty$ and $\dfrac{1}{x0.6^{-x}} \to 0$

$$\lim_{x \to \infty} \dfrac{3}{-(\ln 0.6)x0.6^{-x}} = \boxed{0}$$

29. Answers will vary but should be similar to the following:
    A limit expression with the indeterminate form
    $$\lim_{x \to \infty} \dfrac{f(x)}{g(x)} = \dfrac{\infty}{\infty}$$
    can also be written as
    $$\lim_{x \to \infty} \dfrac{\dfrac{1}{g(x)}}{\dfrac{1}{f(x)}} = \dfrac{0}{0}$$
    or as
    $$\lim_{x \to \infty} \dfrac{1}{f(x)} \cdot g(x) = \dfrac{1}{\infty} \cdot \infty = 0 \cdot \infty$$
    The three indeterminate forms are equivalent.

Chapter 3 Review Activities (pages 246–248)

1. $f(x) = 3.9x^2 + 7x - 5$
   $f'(x) = 3.9(2x) + 7(1) - 0$ by the Sum, Constant Multiplier, Power, and Constant Rules
   $\boxed{f'(x) = 7.8x + 7}$

3. $h(x) = e^{-2x} - e^2$
   $\dfrac{dh}{dx} = (e^{-2x})(-2) + 0$ by the Sum, Exponential, Chain, and Constant Rules
   $\boxed{\dfrac{dh}{dx} = -2e^{-2x}}$

5. $g(x) = 4x - 7\ln x$
   $\dfrac{dg}{dx} = 4 - 7\left(\dfrac{1}{x}\right)$ by the Difference, Constant Multiplier, Power, and Natural Log Rules
   $\boxed{\dfrac{dg}{dx} = 4 - \dfrac{7}{x}}$

7. $j(x) = 2(1.7^{3x+4})$
   $j'(x) = 2(\ln 1.7)(1.7^{3x+4})(3)$ by the Constant Multiplier, Chain, and Power Rules

$$\boxed{j'(x) = 6(\ln 1.7)(1.7^{3x+4})}$$

9. $m(x) = 49x^{0.29}e^{0.7x}$

   $m'(x) = 49(0.29x^{0.29-1})e^{0.7x} + 49x^{0.29}e^{0.7x}(0.7)$ by the Product, Constant Multiplier, Power, Exponential, and Chain Rules

   $$\boxed{m'(x) = 14.21x^{-0.71}e^{0.7x} + 34.3x^{0.29}e^{0.7x}}$$

11. $s(t) = \dfrac{\pi^2}{3t+4}$ rewrite as $s(t) = \pi^2(3t+4)^{-1}$

    $s'(t) = -\pi^2(3t+4)^{-2} \cdot 3$ by the Chain, Sum, Constant Multiplier, and Power Rules

    $$\boxed{s'(t) = -3\pi^2(3t+4)^{-2}}$$

    Alternate form: $s'(t) = \dfrac{-3\pi^2}{(3t+4)^2}$

13. a. $f'(x) = 110.156x - 953.72$ million transactions per year gives the rate of change in monthly transactions where x is the number of years since 1996.

    b. $f(12) = \boxed{2998.192 \text{ million transactions}}$

    c. $f'(12) = \boxed{368.152 \text{ million transactions per year}}$

    $\boxed{\text{The number of transactions monthly per ATM was increasing by 368.152 million transactions per year in 2008.}}$

15. $p'(t) = 0.117t^2 - 0.992t + 2.024$

    a. $p'(10) = 3.804$ percentage points per year

    b. $p'(4) = -0.072$ percentage points per year

    c. $p'(12) - p'(10) = 3.164$ percentage points per year

17. a. rewrite $f(x) = \dfrac{44.58}{1+38.7e^{-0.5x}}$ as $f(x) = 44.58(1+38.7e^{-0.5x})^{-1}$

    $\boxed{f'(x) = -44.58(1+38.7e^{-0.5x})^{-2}(38.7(-0.5e^{-0.5x})) \text{ million homes per year gives the rate of change in the number of homes with access to the Internet via cable television where } x \text{ is the number of years since 1997, data from } 0 \le x \le 11.}$

    Alternate form: $f'(x) = \dfrac{-44.58(38.7(-0.5e^{-0.5x}))}{(1+38.7e^{-0.5x})^2}$

    $= \dfrac{862.623e^{-0.5x}}{(1+38.7e^{-0.5x})^2}$

    b. Internet access: $f(10) \approx \boxed{35.360 \text{ million homes}}$

    ROC: $f'(10) \approx \boxed{3.657 \text{ million homes per year}}$

Chapter 3 Review Activities
Solutions to Odd-Numbered Activities

19. a. $f(x) \approx -25{,}438.507 + 17{,}491.307 \ln x$ dollars gives the median family income x years after 1930, data from $17 \leq x \leq 77$.

    $f(g) = -25{,}438.507 + 17{,}491.307 \ln(g - 1930)$ dollars gives the median family income in year g, data from 1947 to 2007.

    b. $f'(g) = \dfrac{17{,}491.307}{g - 1930}$ dollars per year gives the rate of change in median family income in year g, data from 1947 to 2007.

    c. 1996 ROC: $f'(66) \approx \boxed{\$265 \text{ per year}}$

    %ROC: $\dfrac{f'(66)}{f(66)} \cdot 100\% \approx \boxed{0.55\% \text{ per year}}$

    2004 ROC: $f'(74) \approx \boxed{\$236 \text{ per year}}$

    %ROC: $\dfrac{f'(74)}{f(74)} \cdot 100\% \approx \boxed{0.47\% \text{ per year}}$

21. a. $f(t) = (101.51 \ln t + 219.28)(-0.17t^2 + 1.0065t + 58.64)$ million prescriptions gives the number of brand name prescriptions filled and sold by supermarket pharmacies, t years since 1994, information from $1 \leq t \leq 13$.

    b. $f'(t) = \dfrac{101.51}{t}(-0.17t^2 + 1.0065t + 58.64) + (101.51 \ln t + 219.28)(-0.34t + 1.0065)$ million prescriptions per year gives the rate of change in the number of brand name prescriptions filled and sold by supermarket pharmacies t years since 1994, information from $1 \leq t \leq 13$.

    c. $f'(12) \approx \boxed{-10.58 \text{ million prescriptions per year}}$

23. $\lim\limits_{t \to 0} \dfrac{3.4t}{1 - e^{-1.6t}} = \dfrac{3.4 \cdot 0}{1 - e^{-1.6 \cdot 0}}$ by replacement and limit rules

    $= \dfrac{0}{1 - 1}$ has the indeterminate form $\boxed{\dfrac{0}{0}}$

    $\lim\limits_{t \to 0} \dfrac{3.4t}{1 - e^{-1.6t}} = \lim\limits_{t \to 0} \dfrac{3.4}{1.6 e^{-1.6t}}$ by L'Hôpital's Rule

    $= \dfrac{3.4}{1.6 e^{-1.6 \cdot 0}}$ by replacement and limit rules

    $= \boxed{2.125}$

25. $\lim\limits_{n \to \infty} \dfrac{n^4 + 10n^3}{50n^5 - 65}$ has the indeterminate form $\boxed{\dfrac{\infty}{\infty}}$

    $\lim\limits_{n \to \infty} \dfrac{n^4 + 10n^3}{50n^5 - 65} = \lim\limits_{n \to \infty} \dfrac{4n^3 + 30n^2}{250n^4}$ by L'Hôpital's Rule

    $= \lim\limits_{n \to \infty} \dfrac{12n^2 + 60n}{1000n^3}$ by L'Hôpital's Rule

$$= \lim_{n\to\infty} \frac{24n+60}{3000n^2} \quad \text{by L'Hôpital's Rule}$$

$$= \lim_{n\to\infty} \frac{24}{3000n} \quad \text{by L'Hôpital's Rule}$$

$$= \boxed{0}$$

The first three applications of L'Hôpital's Rule each leads to the indeterminate form $\frac{\infty}{\infty}$.

Chapter 4 Analyzing Change: Applications of Derivatives

Section 4.1 Linearization and Estimates (pages 254–257)

1. a. The rate of change is given in terms of hours and the interval is given in minutes. Convert the interval to hours:

    $$(20 \text{ minutes}) \cdot \frac{1 \text{ hour}}{60 \text{ minutes}} = \frac{1}{3} \text{ hour}$$

    approximated change = (ROC)·(interval)

    $$= \frac{-4 \text{ percentage pts}}{\text{hour}} \cdot \frac{1}{3} \text{ hour}$$

    $$\approx -1.333 \quad \rightarrow \quad \boxed{-1.3 \text{ percentage points}}$$

    (It makes sense to round in context.)

    b. approximated humidity = (initial output) + (approximated change)

    $$\approx 32 - 1.333$$

    $$= 30.667\% \quad \rightarrow \quad \boxed{30.7\%} \quad \text{(It makes sense to round in context.)}$$

3. a. approximated change = (ROC)·(interval)

    $$= \frac{3 \text{ filters}}{\text{day}} \cdot 7 \text{ days}$$

    $$= \boxed{21 \text{ filters}}$$

    b. approximated speed = (initial output) + (approximated change)

    $$= 100 + 21$$

    $$= \boxed{121 \text{ filters}}$$

5. a. $f_L(x) = 17 + 4.6\ x - 3$

    b. $f_L(3.5) = \boxed{19.3}$

7. a. $f_L(x) = 5 - 0.3\ x - 10$

    b. $f_L(10.4) = \boxed{4.88}$

9. a. $R_L(x) = 141 + 38(x - 1.5)$ billion dollars gives the revenue from new car sales when x million dollars are spent on associated advertising expenditures, data from $1.2 \le x \le 6.5$.

    b. $R_L(1.6) = \boxed{\$144.8 \text{ billion}}$

    c. $R_L(2) = \boxed{\$160 \text{ billion}}$

11. a. $g_L(x) = 38.3 - 4.9\ x - 18$ thousand metric tons gives the CFC production where x is the number of years since 1990.

$g_L(19) = 33.4$ thousand metric tons

b. $g_{L2}(x) = 42.2 - 2.9\ x - 17$ thousand metric tons

$g_{L2}(19) = 36.4$ thousand metric tons

c. The estimate in part b is the closer estimate, because $37.7 - 36.4 = 1.3$ while $37.7 - 33.4 = 4.3$.

13. a. future value: $F(10) \approx \$393.156$ thousand

ROC: $F'(t) = 120 \cdot \ln 1.126 \cdot (1.126^t)$

$F'(10) \approx \$46.656$ thousand per year

b. $F_L(t) \approx 393.156 + 46.656\ t - 10$ thousand dollars

c. $F_L(10.5) = 416.484$ thousand dollars

15. (The answer key in the textbook uses a less accurately drawn tangent line than the one used here.)

a.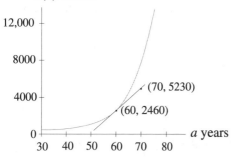

Using $(60, 2460)$ and $(70, 5230)$ estimated from the graph,

$$\text{slope} = \frac{5230 - 2460}{70 - 60}$$

$= 277$ dollars per year

b. $C_L(a) = 2460 + 277\ a - 60$

c. $C_L(63) = \$3291$

17. a. Because f is a positive exponential function, it is increasing and concave up for $w > 0$.

b. Answers may vary but should be similar to the following:

A linearization follows a line tangent to the curve. Because f is concave up, the tangent line at any point on f will lie beneath the curve except at the point of tangency. So a linearization will underestimate the function output values.

19. a.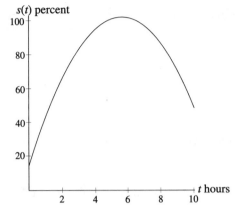

Because s is a negative quadratic function, it is concave down over all t. A graph of s over $0 \le t \le 10$ shows s increasing for $t < 5.6$ and decreasing for $t > 5.6$.

b. Answers may vary but should be similar to the following:
A linearization follows a line tangent to the curve. Because s is concave down for any t, the tangent line at any point on s will lie above the curve except at the point of tangency. So a linearization of s will always overestimate the function values.

21. Answers may vary but should be similar to the following:
When rates of change are used to approximate change in a function, approximations over shorter intervals are generally better than approximations over longer intervals because the tangent line lies closer to the graph of the function near the point of tangency. (See figure.)

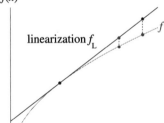

Functions that change concavity and direction may have tangent lines that intersect the function graph at points other than the point of tangency. For such functions, there is a small interval around the point of tangency over which the local linearity principle can be applied. (See figure below.)

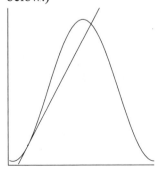

Section 4.2 Relative Extreme Points (pages 264–266)

1. $x = 3$: relative maximum
   The derivative is 0 because the tangent line at $x = 3$ is horizontal.

3. $x = 1$: relative maximum
   The derivative is 0 because the tangent line at $x = 1$ is horizontal.
   $x = 3$ does not present a possibility as a relative minimum because the function is defined by the higher of the two pieces.

5. (The answer key in the lists $x = 5$ instead of $x = 4$ as being the input corresponding to a relative maximum.)
   $x = 1$: relative maximum
   The derivative is 0 because the tangent line at $x = 1$ is horizontal.
   $x = 3$: relative minimum
   The derivative does not exist because the limit of secants from the right does not equal the limit of secants from the left at this point.
   $x = 4$: relative maximum
   The derivative is 0 because the tangent line at $x = 4$ is horizontal.

7. one possible function

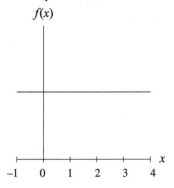

9. i is true because f is increasing for $x < 2$
   ii is true because f is increasing for $x > 2$
   iii is true because f is the tangent at $x = 2$ is horizontal.

11. i is false because f is decreasing for $x < 2$
    ii is true because f is increasing for $x > 2$
    iii is true because f is the tangent at $x = 2$ is horizontal.

13. i is true because f is decreasing for $x < 2$
    ii is true because f is decreasing for $x > 2$
    iii is true because f is the tangent at $x = 2$ is horizontal.

15. i is false because f is increasing for $x < 2$
    ii is true because f is decreasing for $x > 2$
    iii is true because f is the tangent at $x = 2$ is horizontal.

17. one possible function

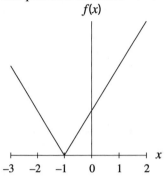

19. one possible function

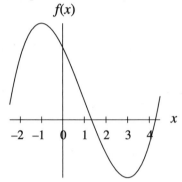

21. a. $f(x) = x^2 + 2.5x - 6$
    $\boxed{f'(x) = 2x + 2.5}$

   b. Solving $f'(x) = 0$ yields $x = -1.25$
      The corresponding output value is $f(-1.25) = -7.5625$ → ≈ -7.563
      (The answer key in the text reports the output value rounded to three decimal places.)
      A graph of f over $-4 < x < 4$ shows a $\boxed{\text{relative minimum at } (-1.25, -7.5625)}$.

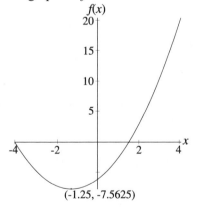

23. a. $h(x) = x^3 - 8x^2 - 6x$
    $\boxed{h'(x) = 3x^2 - 16x - 6}$

b. Solving $h'(x) = 0$ yields two solutions:

$x \approx -0.352$ with corresponding output value $h(-0.352) \approx 1.077$

$x \approx 5.685$ with corresponding output value $h(5.685) \approx -108.929$

A graph of h over $-5 < x < 10$ shows a $\boxed{\text{relative maximum near } (-0.352, 1.077)}$ and a $\boxed{\text{relative minimum near } (5.685, -108.929)}$.

(graph shown on the next page)

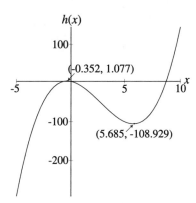

25. a. $f(x) = 12(1.5^x) + 12(0.5^x)$

$\boxed{f'(x) = 12(\ln 1.5)(1.5^x) + 12(\ln 0.5)(0.5^x)}$

b. Solving $f'(x) = 0$ yields $x \approx 0.488$

The corresponding output value is $f(0.488) \approx 23.182$

A graph of f over $-3 < x < 5$ shows a $\boxed{\text{relative minimum near } (0.488, 23.182)}$.

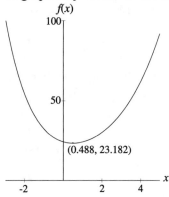

27. a. A graph of $g(x) = 0.04x^3 - 0.88x^2 + 4.81x + 12.11$ shows both a relative maximum and a relative minimum.

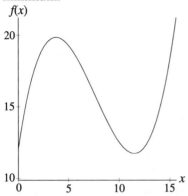

$g'(x) = 0.12x^2 - 1.76x + 4.81$
Solving $g'(x) = 0$ yields two solutions:
$\quad x \approx 3.633$ with corresponding output value $g(3.633) \approx 19.888$
$\quad x \approx 11.034$ with corresponding output value $g(11.034) \approx 11.779$
The function g has a relative maximum near (3.633, 19.888) and a relative minimum near (11.034, 11.779).

b.

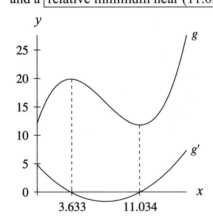

29. a. $f(0) = 123.02$ cfs
$f(11) = 331.305$ cfs

b. $f(h) = -0.865h^3 + 12.05h^2 - 8.95h + 123.02$ cfs
$f'(h) = -2.595h^2 + 24.10h - 8.95$ cfs per hour
Solving $f'(h) = 0$ yields two solutions:
$\quad h \approx 0.388$ hour with corresponding output value $h(0.388) \approx 121.311$
$\quad h \approx 8.900$ hours with corresponding output value $h(8.900) \approx 388.047$
A graph of f over $-1 \leq h \leq 10$ shows a relative minimum near (0.388 hour, 121.311 cfs) and a relative maximum near (8.900 hours, 388.047 cfs).

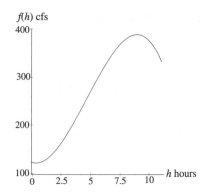

31. The models which could have relative maxima or minima are the quadratic, cubic, and sine models. These are the only models that exhibit a change in direction.

Section 4.3 Absolute Extreme Points (pages 271–273)

1. $x = 1$ and $x = 5$: absolute minimum
   Both inputs have the same minimum value of $y = 2$.
   The derivative does not exist because derivatives are not defined at endpoints.
   $x = 3$: absolute maximum
   The derivative is 0 because the tangent line at $x = 3$ is horizontal.

3. $x = 5.5$: absolute maximum
   The derivative does not exist because derivatives are not defined at endpoints.
   $x = 1$ does not correspond to an absolute maximum because, even thought it corresponds to a relative maximum, there are input values for which the output value is greater.
   $x = 3$ does not correspond to an absolute minimum because the function never actually reaches its right-hand limit at for that input.

5. $x = 1$: absolute maximum
   The derivative is 0 because the tangent line at $x = 1$ is horizontal.
   $x = 6$: absolute minimum
   The derivative does not exist because derivatives are not defined at endpoints.
   $x = 3$ does not correspond to an absolute minimum because even though it corresponds to a relative minimum, there are other input values for which the output values are less.
   $x = 4$ does not correspond to an absolute maximum because even though it corresponds to a relative maximum, there are other input values for which the output values are greater.

7. Candidates for the absolute extreme points of $f(x) = x^2 + 2.5x - 6$ over $-5 \le x \le 5$
   are found among the following points:
   $(-5, 6.5)$ the left endpoint ($f(-5) = 6.5$)
   $(-1.25, -7.5625)$ a relative minimum point (See solution to Section 4.2 Activity 21a.)
   $(5, 31.5)$ the right endpoint ($f(5) = 31.5$)
   Comparing output values at these points shows an absolute minimum of -7.5625 at $x = -1.25$
   and an absolute maximum of 31.5 at $x = 5$.

9. Candidates for the absolute extreme points of $h(x) = x^3 - 8x^2 - 6x$ over $-2 \leq x \leq 10$ are found among the following points:
   - $(-2, -28)$ the left endpoint ($h(-2) = -28$)
   - $(-0.352, 1.077)$ a relative maximum point (See solution to Section 4.2 Activity 23a.)
   - $(5.685, -108.929)$ a relative minimum point (See solution to Section 4.2 Activity 23a.)
   - $(10, 140)$ the right endpoint ($h(10) = 140$)

   Comparing output values at these points shows an $\boxed{\text{absolute minimum of } -108.929 \text{ at } x \approx 5.685}$ and an $\boxed{\text{absolute maximum of } 140 \text{ at } x = 10}$.

11. Candidates for the absolute extreme points of $f(x) = 12(1.5^x) + 12(0.5^x)$ over $-3 \leq x \leq 5.1$ are found among the following points:
    - $(-3, 99.556)$ the left endpoint ($f(-3) \approx 99.556$)
    - $(0.488, 23.182)$ a relative minimum point (See solution to Section 4.2 Activity 25a.)
    - $(5.1, 95.246)$ the right endpoint ($f(5.1) \approx 95.246$)

    Comparing output values at these points shows an $\boxed{\text{absolute maximum of } 99.556 \text{ at } x = -3}$ and an $\boxed{\text{absolute minimum of } 23.182 \text{ at } x \approx 0.488}$.

13. a. A graph of $g(t) = -0.0065t^4 + 0.49t^3 - 13t^2 + 136.3t - 394$ shows that the absolute maximum over $7 \leq t \leq 25$ occurs at a relative maximum point where $g'(t) = 0$.

    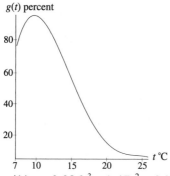

    $g'(t) = -0.026t^3 + 1.47t^2 - 26t + 136.3$ percentage points per degree celsius

    Solving $g'(t) = 0$ yields $t \approx 9.437$ → $\boxed{9.4°C}$

    (Even though it is possible for a cubic to have up to three real roots, g' has only one.)

    The corresponding output value is $g(9.437) \approx 94.781$ → $\boxed{94.8\%}$

    b. The graph of g in the solution to part a, shows that the absolute minimum over $7 \leq t \leq 25$ occurs at the right endpoint: $t = \boxed{25°C}$.

    The absolute minimum value is $g(25) = 5.6875$ → $\boxed{5.7\%}$

15. a. $f(0) = \boxed{123.02 \text{ cfs}}$

    $f(11) = \boxed{331.305 \text{ cfs}}$

    b. $f(h) = -0.865h^3 + 12.05h^2 - 8.95h + 123.02$

    $f'(h) = -2.595h^2 + 24.10h - 895$

    Solving $f'(h) = 0$ yields two solutions:

$h \approx 0.388$ with corresponding output $f(0.388) \approx 121.311$ cfs

$h \approx 8.900$ with corresponding output $f(8.900) \approx 388.047$ cfs

Comparing the output values of the critical points to the output values of the end points found in the solution to part a, shows that over $0 \leq h \leq 11$

the absolute minimum is $\boxed{121.311 \text{ cfs}}$ and

the absolute maximum is $\boxed{388.047 \text{ cfs}}$.

17. (The answer key in the text uses lower case letters to represent these functions. Upper case letters are more commonly used when representing sales, revenue, and profit.)

   a. $S(x) \approx -0.715x^2 + 31.509x - 185.615$ dozen roses gives the number of dozen roses sold by a street vendor when x dollars is the price for a dozen roses, data from $20 to $32.

   b. $R(x) \approx (-0.715x^2 + 31.509x - 185.615) \cdot x$ dollars gives the revenue from the sales of dozen roses by a street vendor when x dollars is the price for a dozen roses, data from $20 to $32.

      Alternate form: $R(x) \approx -0.715x^3 + 31.509x^2 - 185.615x$

   c. (The solution in the answer key in the textbook is rounded incorrectly.)
      Maximum consumer expenditure is the same as maximum producer revenue.

      $R'(x) \approx -2.145x^2 + 63.018x - 185.615$

      Solving $R'(x) = 0$ yields two solutions

      $x \approx 3.321$ → $3.32 which leads to a relative minimum

      $x \approx 26.052$ → $\boxed{\$26.05}$

      The revenue corresponding to $x \approx 26.052$ is greater than the revenue at either endpoint:

      $R(26.052) \approx 3904.533$

      left endpoint: $R(20) \approx 3170.040 < 3904.533$

      right endpoint: $R(32) \approx 2891.201 < 3904.533$

      Alternate solution considering the discreteness of sales:
      If the discrete nature of sales is considered, it is necessary to determine the price that leads to the greatest revenue where $S(x)$ is a whole number.

      $S(26.052) \approx 149.877$

      When 149 dozen roses are sold:
         Solving $S(x) = 149$ yields $x \approx 26.201$ (the other solution is not of interest)
            with revenue $R(26.201) \approx 3903.985$

      When 150 dozen roses are sold:
         Solving $S(x) = 150$ yields $x \approx 26.030$ (the other solution is not of interest)
            with revenue $R(26.030) \approx 3904.522$

      Revenue is higher when the price is $26.03 than when the price is $26.05, because one sale will be lost when the price is raised above $26.03.

   d. Assuming the vendor sells all of the roses he buys, $P(x) = R(x) - 10 \cdot S(x)$ dollars gives profit.

      Alternate form: $P(x) \approx (-0.715x^3 + 31.509x^2 - 185.615x) - 10(-0.715x^2 + 31.509x - 185.615)$

      $\approx -0.715x^3 + 38.661x^2 - 500.707x + 1856.155$

      Solving $P'(x) = 0$ yields two solutions:

      $x \approx 8.462$ → $8.46 which does not make sense in context

      $x \approx 27.576$ → $\boxed{\$27.58}$

      The profit corresponding to $x \approx 27.576$ is greater than the profit at either endpoint:

      $P(27.576) \approx 2450.884$

left endpoint: $P(20) \approx 1585.020 \;<\; 2450.884$

right endpoint: $P(32) \approx 1987.701 \;<\; 2450.884$

Alternate solution considering the discreteness of sales:
If the discrete nature of sales is considered, it is necessary to determine the price that leads to the greatest profit where $S(x)$ is a whole number.

$S(27.576) \approx 139.444$

When 139 dozen roses are sold:
Solving $S(x) = 139$ yields $x \approx 27.632$ (the other solution is not of interest)
with profit $P(27.632) \approx 2450.821$

When 140 dozen roses are sold:
Solving $S(x) = 140$ yields $x \approx 27.506$ (the other solution is not of interest)
with profit $P(27.506) \approx 2450.782$

Profit is higher when the price is $27.63 than when the price is $27.58, because even though lowering the price results in one more sale, all sales bring in less revenue per dozen.

19. a. $s(x) \approx 0.181x^2 - 8.463x + 147.376$ seconds gives the time required for an average athlete to swim 100 meters freestyle at an age of x years, data for ages between 8 and 32 years.

b. Because the function is a concave-up parabola, the relative minimum point is also the absolute minimum point.

$s'(x) \approx 0.362x - 8.463$ seconds per year

Solving $s'(x) = 0$ yields $x \approx 23.368$ → 23 years, 4 months

The minimum value is $s(23.368) \approx 48.498$ → 48.5 seconds.

c. The minimum value on the table is a swim time of 49 seconds at age 24 years.

21. A graph of $y = \dfrac{2x^2 - x + 3}{x^2 + 2}$ over $-50 \le x \le 50$ illustrates the horizontal asymptote at $y = 2$ in both the positive and negative x directions.

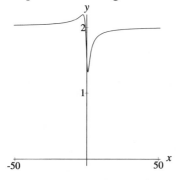

Absolute extrema will occur at critical points.
Solving $y' = 0$ yields the two solutions:

$x \approx -2.732$ with corresponding output $y(-2.732) \approx 2.183$, absolute maximum

$x \approx 0.732$ with corresponding output $y(0.732) \approx 1.317$, absolute minimum

Section 4.4 Inflection Points and Second Derivatives (pages 280–283)

1. (The answer key in the text underestimates the height of the inflection points.)

    a.

    Inflection points can be estimated as the point of greatest slope, $(79, 23)$, and the point of least slope, $(120, 23)$.

    b. The ultimate crude oil production recoverable from Earth was increasing most rapidly in 1979 and was predicted to be decreasing most rapidly in 2020.

3. Figure a shows a graph that crosses the x-axis at $x = b$ and has relative extrema at $x = 0$ and $x = a$. Figure b shows a graph with a relative minimum at $x = b$ and inflection points at $x = 0$ and $x = a$. These points along with the behavior (direction, curvature, and end behavior) of these two graphs suggest that Figure a shows a slope graph of the graph in Figure b.

    Figure c shows a graph that crosses the x-axis at $x = 0$ and $x = a$. These two input values correspond with the relative extrema in Figure a (as well as with the inflection points in Figure b.) Figure c also shows a relative minimum between $x = 0$ and $x = a$. This input corresponds to an inflection point in Figure a. These points along with the general behavior of the graphs in Figures a and c suggest that Figure c shows a slope graph of the graph in Figure a.

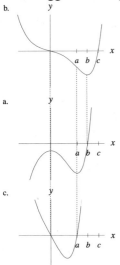

The relationships among the graphs in the three figures are as follows:
- function: Figure b
- derivative: Figure a
- second derivative: Figure c

5. Figure *a* shows a graph that crosses the *x*-axis at $x = -2$ and has a relative minimum at $x = -3$. Figure *b* shows a graph with a relative minimum at $x = -2$ and an inflection point at $x = -3$. These points along with the behavior (direction, curvature, and end behavior) of these two graphs suggests that Figure *a* shows a slope graph of the graph in Figure *b*.

Figure *c* shows an inflection point that corresponds to the same input, $x = -2$, as the relative minimum in Figure *b* (and where the graph in Figure *a* crosses the *x*-axis). Figure *c* also shows a relative minimum at $x = -1$. This input corresponds to a zero in Figure *b*. These points along with the behavior of the graphs in Figures *b* and *c* suggest that Figure *b* shows a slope graph of the graph in Figure *c*.

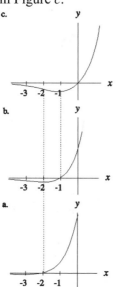

The relationships among the graphs in the three figures are as follows:

function: Figure *c*
derivative: Figure *b*
second derivative: Figure *a*

7. $f(x) = -3x + 7$
$f'(x) = -3$
$f''(x) = 0$

9. $c(u) = 3u^2 - 7u + 5$
$c'(u) = 6u - 7$
$c''(u) = 6$

11. $p(u) = -2.1u^3 + 3.5u^2 + 16$
$p'(u) = -6.3u^2 + 7u$
$p''(u) = -12.6u$

13. $g(t) = 37(1.05^t)$
$g'(t) = 37(\ln 1.05)(1.05^t)$
$g''(t) = 37(\ln 1.05)^2(1.05^t)$

15. $f(x) = 3.2\ln x + 7.1$

$f'(x) = 3.2 \cdot \dfrac{1}{x}$

Rewrite: $\boxed{f'(x) = 3.2x^{-1}}$

$\boxed{f''(x) = -3.2x^{-2}}$

Alternate form: $f''(x) = \dfrac{-3.2}{x^2}$

17. $L(t) = \dfrac{16}{1 + 2.1e^{3.9t}}$

Rewrite as $L(t) = 16(1 + 2.1e^{3.9t})^{-1}$

$\boxed{L'(t) = -16(1 + 2.1e^{3.9t})^{-2}(2.1 \cdot 3.9e^{3.9t})}$ by the chain rule

Alternate form: $L'(t) = -131.04e^{3.9t}(1 + 2.1e^{3.9t})^{-2}$

$= \dfrac{-131.04e^{3.9t}}{(1 + 2.1e^{3.9t})^2}$

$L''(t) = \dfrac{d}{dt}(-131.04e^{3.9t}) \cdot (1 + 2.1e^{3.9t})^{-2} + (-131.04e^{3.9t}) \cdot \dfrac{d}{dt}(1 + 2.1e^{3.9t})^{-2}$ by the product rule

$\boxed{L''(t) = (-131.04 \cdot 3.9e^{3.9t}) \cdot (1 + 2.1e^{3.9t})^{-2} + (-131.04e^{3.9t}) \cdot [-2(1 + 2.1e^{3.9t})^{-3}(2.1 \cdot 3.9e^{3.9t})]}$

by the chain rule: $\dfrac{d}{dt}(1 + 2.1e^{3.9t})^{-2} = \dfrac{d}{du}u^{-2} \cdot \dfrac{d}{dt}u$ where $u = 1 + 2.1e^{3.9t}$

$= -2(1 + 2.1e^{3.9t})^{-3} \cdot (2.1 \cdot 3.9e^{3.9t})$

Alternate form: $L''(t) = -511.056e^{3.9t}(1 + 2.1e^{3.9t})^{-2} + 2146.4352e^{7.8t}(1 + 2.1e^{3.9t})^{-3}$

$= \dfrac{-511.056e^{3.9t}}{(1 + 2.1e^{3.9t})^2} + \dfrac{2146.4352e^{7.8t}}{(1 + 2.1e^{3.9t})^3}$

19. $f(x) = x^3 - 6x^2 + 12x$

$f'(x) = 3x^2 - 12x + 12$

$f''(x) = 6x - 12$

Solving $f''(x) = 0$ yields $\boxed{x = 2}$

21. $f(x) = \dfrac{3.7}{1 + 20.5e^{-0.9x}}$

Rewrite as $f(x) = 3.7(1 + 20.5e^{-0.9x})^{-1}$

$f'(x) = -3.7(1 + 20.5e^{-0.9x})^{-2} \cdot (20.5 \cdot (-0.9e^{-0.9x}))$ by the chain rule

Rewrite as $f'(x) = 68.265e^{-0.9x}(1 + 20.5e^{-0.9x})^{-2}$

Alternate form: $f'(x) = \dfrac{68.265e^{-0.9x}}{(1 + 20.5e^{-0.9x})^2}$

$f''(x) = (68.265 \cdot (-0.9e^{-0.9x})) \cdot (1 + 20.5e^{-0.9x})^{-2} + 68.265e^{-0.9x} \cdot [-2(1 + 20.5e^{-0.9x})^{-3}(20.5 \cdot (-0.9e^{-0.9x}))]$

by the product and chain rules

$$f''(x) = -61.4385e^{-0.9x}(1+20.5e^{-0.9x})^{-2} + 2518.9785e^{-1.8x}(1+20.5e^{-0.9x})^{-3}$$

Alternate form:
$$= \frac{-61.4385e^{-0.9x}}{(1+20.5e^{-0.9x})^2} + \frac{2518.9785e^{-1.8x}}{(1+20.5e^{-0.9x})^3}$$

Solving $f''(x) = 0$ yields $\boxed{x \approx 3.356}$

23. $f(x) = 98(1.2^x) + 120(0.2^x)$
$f'(x) = 98 \cdot \ln 1.2(1.2^x) + 120 \cdot \ln 0.2(0.2^x)$
$f''(x) = 98(\ln 1.2)^2(1.2^x) + 120(\ln 0.2)^2(0.2^x)$
There is no inflection point on f because f'' is always positive.

25. a. $g(x) = 0.04x^3 - 0.88x^2 + 4.81x + 12.11$
$g'(x) = 0.12x^2 - 1.76x + 4.81$
$g''(x) = 0.24x - 1.76$

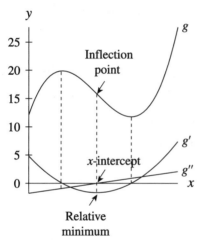

b. The inflection point occurs where the second derivative crosses the *x*-axis.
Solving $g''(x) = 0$ yields $x \approx \boxed{7.333}$

The corresponding output value is $g(7.333) \approx \boxed{15.834}$
The graph of g in the solution of part *a* shows that (7.333, 15.834) is the $\boxed{\text{point of most rapid decline}}$.

27. a. $p(x) = -0.00022x^3 + 0.014x^2 - 0.0033x + 12.236$
$p'(x) = -0.00066x^2 + 0.028x - 0.0033$
$p''(x) = -0.00132x + 0.028$
The figure on the left shows all three graphs. The figure on the right shows the first and second derivative graphs on with a larger scale on the vertical axis so that detail can be seen.

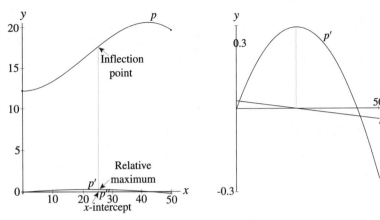

The graph of p shows the point of most rapid increase occurs at the inflection point.
Solving $p''(x)=0$ yields $x \approx 21.212$ → March, 2021

The corresponding output value is $p(21.212) \approx$ 16.366%

The rate of change at that input is $p'(21.212) \approx$ 0.294 percentage points per year

b. The graph of g in the solution to part a shows the point of most rapid decrease occurs at the right endpoint: $x = 50$ → the end of 2050

The corresponding output value is $p(50) \approx$ 19.571%

The rate of change at that input is $p'(50) \approx$ −0.253 percentage points per year

29. a. $p(x) = 0.0987x^4 - 2.1729x^3 + 17.027x^2 - 55.023x + 72.133$ dollars
$p'(x) = 0.3948x^3 - 6.5187x^2 + 34.054x - 55.023$ dollars per year
$p''(x) = 1.1844x^2 - 13.0374x + 34.054$
Solving $p''(x) = 0$ yields two solutions:

$x \approx 4.263$ → April 2005
The corresponding output value is $p(4.263) \approx 11.262$

The inflection point is (4.263, 11.262) and has ROC $p'(4.263) \approx$ 2.270

$x \approx 6.745$ → September 2007
The with corresponding output $p(6.745) \approx 13.151$

The inflection point is (6.745, 13.151) and has ROC $p'(6.745) \approx$ −0.747

b. Between 2004 and 2010, the average price for 1000 cubic feet of natural gas was increasing most rapidly in April 2005 by $2.270 per year. At that time the price for 1000 cubic feet of natural gas was $11.26.
Between 2004 and 2010, the average price for 1000 cubic feet of natural gas was decreasing most rapidly in September 2007 by $0.747 per year. At that time the price for 1000 cubic feet of natural gas was $13.15.

31. a. $P(x) = 6 + 62.7(1 + 38.7e^{-0.258x})^{-1}$ percent
$P'(x) = -62.7(1 + 38.7e^{-0.258x})^{-2}(38.7(-0.258e^{-0.258x}))$
$\approx 626.034e^{-0.258x}(1 + 38.7e^{-0.258x})^{-2}$ percentage points per year

$P'(x) \approx 626.034(-0.258e^{-0.258x})(1+38.7e^{-0.258x})^{-2} + 626.034e^{-0.258x}[-2(1+38.7e^{-0.258x})^{-3}(38.7(-0.258e^{-0.258x}))]$

$\approx -161e^{-0.258x}(1+38.7e^{-0.258x})^{-2} + 12{,}501.407e^{-0.516x}(1+38.7e^{-0.258x})^{-3}$

Solving $P''(x) = 0$ yields $x \approx 14.170$ → March 1985

A graph of P shows that this inflection point is the point of most rapid increase.

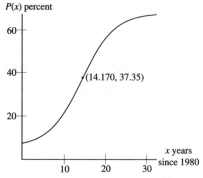

b. Percentage of households: $P(14.170) \approx$ 37.350%

ROC: $P'(14.170) \approx$ 4.044 percentage points per year

c. Between 1970 and 2002, the percentage of households with TVs whose owners subscribed to cable was increasing most rapidly in March 1985 – by 4.044 percentage points per year. At that time, the owners of 37.35 percent of households with TVs subscribed to cable.

33. a. $f(x) \approx \dfrac{10{,}111.102}{1+1153.222e^{-0.728x}}$ hours gives the total number of labor hours spent on a construction job where x is the number of weeks since the start of construction, data from 1 to 19 weeks.

b. $f(x) \approx 10{,}111.102(1+1153.222e^{-0.728x})^{-1}$

$f'(x) \approx -10{,}111.102(1+1153.222e^{-0.728x})^{-2}(1153.222(-0.728e^{-0.728x}))$

$f'(x) \approx 8{,}488{,}330.434e^{-0.728x}(1+1153.222e^{-0.728x})^{-2}$ hours per week

c. $f''(x) \approx 8{,}488{,}330.434(-0.728e^{-0.728x})(1+1153.222e^{-0.728x})^{-2}$

$+ 8{,}488{,}330.434e^{-0.728x}[-2(1+1153.222e^{-0.728x})^{-3}(1153.222(-0.728e^{-0.728x}))]$

$\approx -7{,}125{,}927{,}058e^{-0.728x}(1+1153.222e^{-0.728x})^{-2} + 14{,}251{,}854{,}116e^{-1.456x}(1+1153.222e^{-0.728x})^{-3}$

Solving $f''(x) = 0$ yields $x \approx$ 9.685 weeks

The corresponding output value is $f(9.685) \approx$ 5,055.551 labor hours

A graph of f shows that this inflection point is the point of most rapid increase.

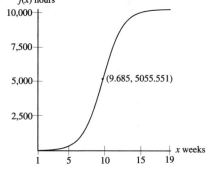

d. Answers may vary but should be similar to the following:
The manager should schedule the second job to begin approximately 10 weeks after the start of the first job begins.

35. a. Comparing differences between consecutive output values in the table shows the interval from $\boxed{1990 \text{ to } 1995}$ has the least increase.
averageROC $= \dfrac{122-115}{1995-1990} = \boxed{1.4 \text{ million tons per year}}$

b. $g(x) = 0.008x^3 - 0.324x^2 + 5.814x + 79.881$ million tons gives the garbage taken yearly to a landfill where x is the number of years since 1980, data from 1980 and 2010.

c. $g'(x) = 0.024x^2 - 0.648x + 5.814$ million tons per year
$g''(x) = 0.048x - 0.648$

Solving $g''(x) = 0$ yields $x = 13.5$ → $\boxed{\text{July 1994}}$
A graph of g shows that the inflection point is the point of slowest increase.

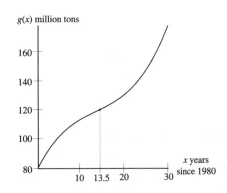

$\boxed{\text{The point of slowest increase occurs at the input value where there is a minimum on the first derivative or at the point where the second derivative is 0.}}$

d. Minimum ROC was found in the solution to part c: $\boxed{1994}$.
ROC: $g'(13.5) = \boxed{1.44 \text{ million tons per year}}$

37. Answers will vary but should include the following information:
When the second derivative of a function is positive, the first derivative of that function is increasing, and the function itself is concave up. Three possible functions are shown in the figures below.

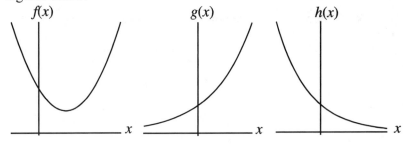

Section 4.5 Marginal Analysis (pages 288–290)

1. At a production level of 500 units, cost is increasing by $17 per unit.

3. When weekly sales are 500 units, revenue is increasing by $10 per unit and cost is increasing by $13 per unit.

5. a. $\dfrac{\text{output units}}{\text{input units}} = \dfrac{\text{\sout{million} dollars}}{\text{\sout{million} units}}$

   $= \boxed{\text{dollars per unit}}$

   b. $\boxed{\text{At a production level of 40 million units, cost is increasing by \$0.20 per unit.}}$

7. a. $\dfrac{\text{output units}}{\text{input units}} = \dfrac{\text{million dollars}}{\text{thousand units}}$

   $= \dfrac{1000 \ \text{\sout{thousand}} \ \text{dollars}}{\text{\sout{thousand}} \ \text{units}}$ (numerator conversion: $1{,}000{,}000 = 1000 \cdot 1000$)

   $= \boxed{\text{thousand dollars per unit}}$

   b. 0.02 thousand dollars per unit $= 0.02 \cdot 1000$ dollars per unit
   $= 20$ dollars per unit

   $\boxed{\text{When 50,000 units are sold, revenue is increasing by \$20 per unit.}}$

9. a. $\dfrac{\text{output units}}{\text{input units}} = \dfrac{\text{hundred dollars}}{\text{thousand units}}$

   $= \dfrac{100 \ \text{dollars}}{1000 \ \text{units}}$

   $= \dfrac{\tfrac{1}{10} \ \text{dollars}}{\text{unit}}$ $\rightarrow$ $\boxed{\text{tenths of a dollar per unit}}$

   Alternate answer: tens of cents per unit

   b. 3 tens of cents per unit $= 3 \cdot 10$ cents per unit
   $= 30$ cents per unit

   $\boxed{\text{When 16 thousand units are sold, revenue is increasing by 30¢ per unit.}}$

11. Answers will vary but might include one or more of the following:
    1) increase the price of the t-shirt,
    2) find a way to lower the cost to the fraternity for the t-shirts, or
    3) stop selling the t-shirts altogether.

13. a. $R(x)$ thousand dollars

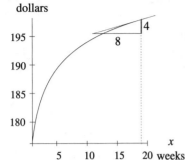

Using rise and run estimated for a tangent line drawn on the graph,
$$R'(19) \approx \frac{4}{8} = \boxed{0.5 \text{ thousand dollars per week}}$$
Using the tangent line at $x = 19$ to approximate revenue at 21 weeks,
$$R(21) \approx R(19) + 2 \cdot R'(19)$$
$$= 199 + 2(0.5)$$
$$= \boxed{\$200 \text{ thousand}}$$

b. $R(21) \approx \boxed{\$198.523 \text{ thousand}}$

15. a. $c'(p) = 0.48p^2 - 17.4p + 172$ dollars per unit
5 units: $c'(5) = \boxed{97 \text{ dollars per unit}}$
20 units: $c'(16) = \boxed{16 \text{ dollars per unit}}$
30 units: $c'(82) = \boxed{82 \text{ dollars per unit}}$

b. 6th unit: $c(5+1) - c(5) = \boxed{\$90.86}$
21st unit: $c(20+1) - c(20) = \boxed{\$17.06}$
31st unit: $c(30+1) - c(30) = \boxed{\$87.86}$

c. Answers will vary but should include the following information:
The marginal cost is found using a linearization of the cost function. The cost function is concave down at $p = 5$ so the marginal cost is greater than the cost function. The cost function is concave up at $p = 20$ and $p = 30$ so the marginal cost is less than the cost function.

17. a. $R(x) = -12.16x^2 + 254.28x - 105.6$ dollars gives the revenue on Friday night from the sales of large one-topping pizzas priced at x dollars, based on pizza costs between $9.25 and $14.25.

b. $R'(x) = -24.32x + 254.28$ dollars per dollar (revenue dollars per price dollar)
$9.25: $R'(9.25) = \boxed{29.32 \text{ dollars per dollar}}$
$11.50: $R'(11.5) = \boxed{-25.40 \text{ dollars per dollar}}$

c. $9.25 to $10.25: $R(10.25) - R(9.25) = \boxed{\$17.16}$
$11.50 to $12.50: $R(12.50) - R(11.50) = \boxed{-\$37.56}$

d. Answers will vary but should include the following information:
The marginal price is found using a linearization of the revenue function. The revenue function is concave down with a maximum value. When $R'(x) > 0$, the linearization overestimates the change because the tangent is above $R(x)$. When $R'(x) < 0$, the linearization underestimates the change because the tangent is above $R(x)$.

Section 4.6 Optimization of Constructed Functions (pages 298–303)

Even though activities in this section can be worked in alternate forms, the optimal answer should be the same.

1. a. output to optimize: area, A square feet
input: width (perpendicular to house), x feet
length, y feet

   b.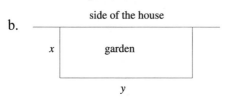

   c. perimeter (3 sides) = $2x + y$
   fencing for 3 sides = 60 feet
   fencing = perimeter (3 sides)
   $$60 = 2x + y$$
   Solving $60 = 2x + y$ for y yields $y = 60 - 2x$
   area of a rectangle: $A = xy$

   $\boxed{A(x) = x(60 - 2x)}$ square feet gives the area of a garden with width x feet and 60 feet of fencing on three sides.

   Alternate form: $A(x) = 60x - 2x^2$

   d. Solving $A'(x) = 0$ yields $x = 15$ feet (width)
   So $y = 60 - 2 \cdot 15 = 30$ feet (length)
   The maximum area is $A(15) = \boxed{450 \text{ square feet}}$
   Because A is a quadratic function with a negative coefficient (–2) on x^2, it's graph is a concave down parabola. The relative extreme must be a maximum.

3. a. output to optimize: area, A square feet
input: width (perpendicular to house), x feet
length, y feet

   b.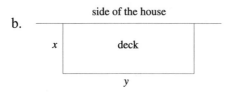

   c. perimeter (3 sides) = $2x + y$
   railing + steps = $32 + 3 = 35$

railing + steps = perimeter (3 sides)
$$35 = 2x + y$$
Solving $35 = 2x + y$ for y yields $y = 35 - 2x$
area of a rectangle: $A = xy$

$\boxed{A(x) = x(35 - 2x) \text{ square feet gives the area of a deck with width } x \text{ and 32 feet of railing along three sides plus a 3-foot wide stairway entrance.}}$

Alternate form: $A(x) = 35x - 2x^2$

d. Solving $A'(x) = 0$ yields $x = \boxed{8.75 \text{ feet}}$

So $y = 35 - 2 \cdot 8.75 = \boxed{17.5 \text{ feet}}$

The maximum area is $A(8.75) = 153.125$ square feet

Because A is a quadratic function with a negative coefficient (–2) on x^2, it's graph is a concave down parabola. The relative extreme must be a maximum.

5. a. output to optimize: area, A square feet
input: width, x feet
height of rectangle, y feet

b.

c. perimeter (semicircle + 3 sides of rectangle) = $0.5\pi x + x + 2y$
outside perimeter as stated = 20 feet
stated perimeter = semicircle + 3 sides
$$20 = 0.5\pi x + x + 2y$$
Solving $20 = 0.5\pi x + x + 2y$ for y yields $y = \dfrac{20 - (0.5\pi x + x)}{2}$

Alternate form: $y = 10 - (0.25\pi + 0.5)x$

area (semicircle + rectangle): $A = 0.5\pi(0.5x)^2 + xy$

$\boxed{A(x) = 0.5\pi(0.5x)^2 + x \cdot \dfrac{20 - (0.5\pi x + x)}{2} \text{ square feet gives the area of a Norman window with perimeter 20 feet and width } x.}$

Alternate form: $A(x) = 0.125\pi x^2 + 10x - (0.25\pi + 0.5)x^2$
$$= 10x - (0.125\pi + 0.5)x^2$$

d. Solving $A'(x) = 0$ yields $x \approx \boxed{5.601 \text{ feet}}$

So $y \approx \dfrac{20 - (0.5\pi \cdot 5.601 + 5.601)}{2} \approx \boxed{2.800 \text{ feet}}$

The maximum area is $A(5.601) \approx 28.005$ square feet.

Because A is a quadratic function with a negative coefficient $(-0.125\pi - 0.5)$ on x^2, it's graph is a concave down parabola. The relative extreme must be a maximum.

7. Step 1: identify variable quantities
   output to optimize: combined areas of top and base: A square inches
   input: one edge of bottom square, x inches   (perimeter of square = $4x$)
       circumference of top, c inches

Step 2: sketch

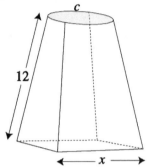

Step 3: construct a model
   length of pieces (4 upright edges + 4 base edges + top circle) = $4 \cdot 12 + 4x + c = 48 + 4x + c$
   total length of wire frame = 9 feet = 108 inches
       total length = length of pieces
       $108 = 48 + 4x + c$
   Solving $108 = 48 + 4x + c$ for c yields $c = 60 - 4x$
   area of a square with side length x: $A_s = x^2$
   area of a circle with circumference c is calculated as:
       The circumference c of a circle is related to its radius r by $c = 2\pi r$.
       Solving $c = 2\pi r$ for r yields $r = \dfrac{c}{2\pi}$.
       The area a of a circle is related to its radius r by $a = \pi r^2$.
   area of a circle with circumference $60 - 4x$: $A_c = \pi r^2$

   $$= \pi \left(\dfrac{c}{2\pi}\right)^2$$

   $$= \pi \left(\dfrac{c}{2\pi}\right)^2$$

   $$= \pi \left(\dfrac{60 - 4x}{2\pi}\right)^2$$

(The solution to Activity 8 is identical until this line.)

$A(x) = x^2 + \dfrac{(60 - 4x)^2}{4\pi}$ square inches gives the combined areas of the top and bottom of the frame where x inches represents the length of one edge of the bottom.

Alternate form: $A(x) = x^2 + \dfrac{3600 - 480x + 16x^2}{4\pi}$

$= \dfrac{900}{\pi} - \dfrac{120}{\pi}x + \left(1 + \dfrac{4}{\pi}\right)x^2$

Step 4: optimize

Solving $A'(x) = 0$ yields $x \approx 8.401$

perimeter of the square: $4x \approx 33.606$ → $\boxed{33.6 \text{ inches}}$ (or 2.8 feet)

circumference of the circle: $c = 60 - 4x \approx 26.394$ → $\boxed{26.4 \text{ inches}}$ (or 2.2 feet)

four upright edges: $\boxed{\text{four 12-inch pieces}}$

Because A is a quadratic function with a positive coefficient $\left(1 + \dfrac{4}{\pi}\right)$ on x^2, it's graph is a concave up parabola. The relative extreme must be a minimum.

9. a. Step 1: identify variable quantities
   output to optimize: area of the garden: A square feet
   input: length of the wood fence, x linear feet
   other dimension of the garden, y linear feet

   Step 2: sketch

   [sketch of rectangle: left side labeled x, wire fence @ $2 per linear foot on top, bottom, and left sides; wood privacy fence @ $6 per linear foot on right side; bottom labeled y]

   Step 3: construct a model
   length of wood fence (1 side) = x feet
   cost of wood fence (1 side) = $6x$ dollars
   length of wire fence (3 sides) = $x + 2y$ feet
   cost of wire fence = $2(x + 2y)$ dollars
   total cost = $320

   total cost = cost of wood fence + cost of wire fence
   $320 = 6x + 2(x + 2y)$
   $320 = 8x + 4y$

   Solving $320 = 8x + 4y$ for y yields $y = \dfrac{320 - 8x}{4} = 80 - 2x$

   area of the rectangular garden: $A = xy$
   $= x(80 - 2x)$

   $A(x) = x(80 - 2x)$ square feet gives the area of the garden surrounded by wood and wire fencing costing $320 when y is the length of wood fencing.

   Alternate form: $A(x) = 80x - 2x^2$

b. Step 4: optimize

Solving $A'(x) = 0$ yields $x = 20$

width of the garden: $x = \boxed{20 \text{ feet}}$

length of the garden: $y = 80 - 2x = \boxed{40 \text{ feet}}$

area of the garden: $A = xy = \boxed{800 \text{ square feet}}$

Because A is quadratic with a negative (–2) coefficient on x^2, its graph is a concave down parabola. The relative extreme must be a maximum.

c. length of wood fence: $x = \boxed{20 \text{ linear feet}}$

length of wire fence: $x + 2y = \boxed{100 \text{ linear feet}}$

11. (The activity should list the units of measure for the x and y axes as miles.)
In order to use standard notation for the cost function, we chose to change the name of the point on the horizontal axis to P instead of C. We use C to represent the cost function.)

a. Step 1: identify variable quantities

output to optimize: overland transportation costs: C dollars
input: coordinates of landing point, x miles
y miles

Step 2: sketch

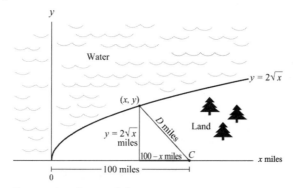

Step 3: construct a model

landing point: $(x, y) = (x, 2\sqrt{x}\,)$

point P (the hut): (100, 0)

distance between points (x_a, y_a) and (x_b, y_b): $d = \sqrt{(x_a - x_b)^2 + (y_a - y_b)^2}$

distance between landing point and hut: $d = \sqrt{(x - 100)^2 + (2\sqrt{x} - 0)^2}$ miles

cost to transport from landing point to hut: $C = 10d$

$= 10\sqrt{(x - 100)^2 + 4x}$ dollars

$C(x) = 10\sqrt{(x - 100)^2 + 4x}$ dollars gives the cost to transport goods overland from the beach to the hut when the landing point is on a line x miles east of the point (0, 0).

Step 4: optimize

Solving $C'(x) = 0$ yields $x = 98$ miles

$(x, y) = (98, 2\sqrt{98}) \quad \rightarrow \quad \approx \boxed{(98 \text{ miles}, 19.799 \text{ miles})}$

A graph of C over $0 < x < 200$ illustrates that the solution corresponds to a relative minimum.

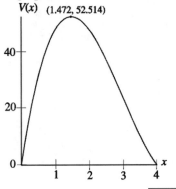

b. $C(98) \approx 198.997$ → $\boxed{\$199.00}$

13. Step 1: identify variable quantities
    output to optimize: cost of run walls and fencing, C dollars
    input quantities: length of cinder block sides, b feet
    length of chain link side, l feet

    Step 2: sketch

    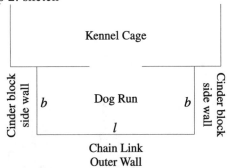

    Step 3: construct a model
    area of the dog run = (length of cinder block wall)(length of chain link fence)
    $$120 = bl$$
    Solving $120 = bl$ for b yields $b = \dfrac{120}{l}$

    square footage of cinder block walls = 2 walls b feet long 7 feet high
    $$= 14b$$
    $$= 14\left(\dfrac{120}{l}\right) \text{ square feet}$$

    cost of cinder block walls = \$0.50 per square foot (area of walls)
    $$= 0.50 \cdot 14\left(\dfrac{120}{l}\right)$$
    $$= \dfrac{840}{l} \text{ dollars}$$

    square footage of chain link fencing = l feet long 7 feet high
    $$= 7l \text{ square feet}$$

cost of chain link fencing = $2.75 per square foot (area of fencing)
$$= 2.75 \cdot 7l$$
$$= 19.25l \text{ dollars}$$
cost of walls and fence = (cost of walls) + (cost of fencing)
$$C = \frac{840}{l} + 19.25l$$

$C(l) = \frac{840}{l} + 19.25l$ dollars gives the cost of the walls and fencing for a 7-foot high, 120 square foot dog run when l feet of chain link fencing is used along one side.

Step 4: optimize
Solving $C'(l) = 0$ yields $l \approx 6.606$
A graph of C over $0 \le l \le 50$ illustrates that the solution is a relative minimum of C.

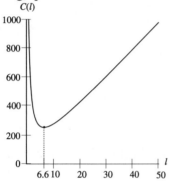

length of chain link fencing: $\boxed{6.6 \text{ feet}}$

length of cinder block wall: $\frac{120}{l} \approx 18.166 \rightarrow \boxed{18.2 \text{ feet}}$

15. Step 1: identify variable quantities
output to optimize: cost of materials, C dollars
input quantities: length of back (display board), b feet
length of sides (gathered fabric), s feet

Step 2: sketch

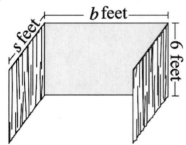

Step 3: construct a model
area of booth = (length of side)(length of back)
$$300 = sb$$
Solving $300 = sb$ for s yields $s = \frac{300}{b}$

square footage of display board = b feet long · 6 feet high
$$= 6b \text{ square feet}$$
cost of display board = \$30 per square foot · (area of display board)
$$= 30 \cdot 6b$$
$$= 18b \text{ dollars}$$
square footage of fabric = 2 sides · $2s$ feet long · 6 feet high
$$= 24s \text{ square feet}$$
$$= 24\left(\frac{300}{b}\right) \text{ square feet}$$
cost of fabric = \$2 per square foot · (area of fencing)
$$= 2 \cdot 24\left(\frac{300}{b}\right)$$
$$= \frac{14{,}400}{b} \text{ dollars}$$
cost of booth = (cost of display board) + (cost of fabric)
$$C = 180b + \frac{14{,}400}{b}$$

$C(b) = 180b + \dfrac{14{,}400}{b}$ dollars gives the cost of a 300 square foot booth, when the back of the booth is b feet wide.

Step 4: optimize

Solving $C'(b) = 0$ yields $b \approx 8.944$

A graph of C over $0 \le b \le 50$ illustrates that the solution is a relative minimum of C.

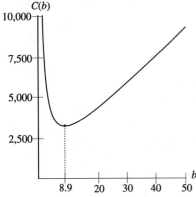

width of booth: $\boxed{8.9 \text{ feet}}$

length of side wall: $\dfrac{300}{b} \approx 33.541 \rightarrow \boxed{33.5 \text{ feet}}$

17. Step 1: identify variable quantities
output to be optimized: cost of order and store cases, C dollars
input: cases needed per order (order size): x cases per order
(Step 2 is skipped because the problem is not geometric.)
Step 3: construct a model (during solutions part a, b, and c)

a. cases needed annually: $\dfrac{42{,}000 \text{ packets}}{1000 \text{ packets per case}} = 42$ cases

orders needed annually: $y = \dfrac{42 \text{ cases}}{x \text{ cases per order}}$

$$\boxed{y = \dfrac{42}{x} \text{ orders}}$$

annual cost associated with ordering = (cost per order)(number of orders needed)

$$C_a = (\$12 \text{ per order}) \cdot \left(\dfrac{42}{x} \text{ orders}\right)$$

$$\boxed{C_a = \dfrac{504}{x} \text{ dollars}}$$

b. average number of cases being stored: $\dfrac{x}{2}$ cases

annual cost associated with storage = (cost per case)(number of cases being stored)

$$C_s = (\$4 \text{ per order}) \cdot \left(\dfrac{x}{2} \text{ cases}\right)$$

$$\boxed{C_s = 2x \text{ dollars}}$$

c. cost to order and store cases = (cost to order) + (cost to store)

$$C = C_a + C_s$$

$$C = \dfrac{504}{x} + 2x$$

$C(x) = \dfrac{504}{x} + 2x$ dollars gives the combined ordering and storage costs when x cases are ordered at a time.

d. Step 4: optimize

Solving $C'(x) = 0$ yields $x \approx 15.875$

A graph of C illustrates that C has a relative minimum at $x \approx 15.875$.

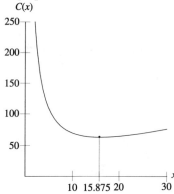

The context requires both order size x and number of orders y to be whole numbers.

orders needed annually: $y = \dfrac{42}{x} \approx 2.646 \rightarrow \boxed{3 \text{ orders}}$

cost associated with three orders = $12 \cdot 3 = \$36$

Two candidates for order size come from the integers on either side of $x \approx 15.875$.

$x = 16$ cases   Orders would be of different sizes: 16 cases, 16 cases, 10 cases.
  storage cost associated with 16 cases = $4 \cdot 8 = \$32$
   (No assumption of pro-rating storage costs for smaller orders is given in the activity.)
  total ordering and storage costs = $36 + 32 = \$68$

$x = 15$ cases   Orders would be of different sizes: 15 cases, 15 cases, 12 cases.
  storage cost associated with 15 cases = $4 \cdot 7.5 = \$30$
  total ordering and storage costs for 15 cases = $36 + 30 = \$66$

One other candidate for order size comes from the number of orders needed: when $y = 3$,

$x = 14$ cases   Orders would be the same size: $\boxed{14 \text{ cases, } 14 \text{ cases, } 14 \text{ cases}}$.
  storage cost associated with 14 cases = $4 \cdot 7 = \$28$
  total ordering and storage costs for 14 cases = $36 + 28 = \boxed{\$64}$

Ordering 14 cases each time is the least costly way to place the orders.

19. (The activity should give storage costs as $1 per thousand tins instead of $1 per tin.)

  a. Step 1: identify variable quantities
    output to be optimized: cost to setup production and store tins, C dollars
    input: tins needed per production run: x million tins per run
  (Step 2 is skipped because the problem is not geometric.)
  Step 3: construct a model
    tins needed annually: 1.7 million tins
    runs needed annually: $y = \dfrac{1.7 \text{ million tins}}{x \text{ million tins per run}}$

    $= \dfrac{1.7}{x}$ runs

    annual cost associated with setup = (cost to setup)(number of runs)

    $$C_a = (\$1300 \text{ per run}) \cdot \left(\dfrac{1.7}{x} \text{ runs} \right)$$

    $$= \dfrac{2210}{x} \text{ dollars}$$

    average number of tins being stored: $\dfrac{x}{2}$ million tins

    annual cost associated with storage = (cost per tin)(number of tins being stored)

    $$C_s = (\$1 \text{ per thousand tin}) \cdot \left(\dfrac{x}{2} \text{ million tins} \right)$$

    $$= \dfrac{x}{2} \text{ thousand dollars}$$

    $$= 500x \text{ dollars}$$

    cost to setup production and store tins = (cost to setup runs) + (cost to store)

    $$C = C_a + C_s$$

    $$= \dfrac{2210}{x} + 500x$$

$C(x) = \dfrac{2210}{x} + 500x$ dollars gives the combined setup and storage costs when x million tins are produced in a run.

Step 4: optimize

Solving $C'(x) = 0$ yields $x \approx 2.102$

A graph of C illustrates that C has a relative minimum at $x \approx 2.102$.

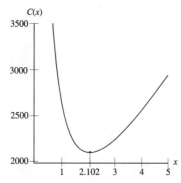

Because the input ($x \approx 2.102$ million tins) corresponding to the relative minimum of C is larger than the necessary number of tins (1.7 million tins), a single run of 1.7 million tins will minimize setup and storage.

b. One run of 1.7 million tins once a year

Alternate solution (assuming that tins can be stored for more than one year)
One run of 2.102 million tins once every 1.237 years (or 1 year, 2 months, 25 days)

21. a. Step 1: identify variable quantities
output to be optimized: profit, P
input: number of one dollar increases, x
(Step 2 is skipped because the problem is not geometric.)
Step 3 (during parts a, b, and c): construct a model
price after x increases = $10 + x$
daily sales after x increases = $20 - 2x$
revenue = (price)(number sold)
$$R = (10 + x)(20 - 2x)$$
$R(x) = (10 + x)(20 - 2x)$ dollars gives revenue when $x + 10$ dollars is the price of the necklace.

Alternate form: $R(x) = 200 - 2x^2$

b. cost = (cost per necklace)(number made)   assume number made = number sold
$$C = 6(20 - 2x)$$
$C(x) = 6(20 - 2x)$ dollars gives the cost of making necklaces $20 - 2x$ necklaces.

Alternate form: $C(x) = 120 - 12x$

c. profit = revenue − cost
$$P(x) = R(x) - C(x)$$
$P(x) = (10 + x)(20 - 2x) - 6(20 - 2x)$ dollars gives profit when $x + 10$ dollars is the price of the necklace.

Alternate form: $P(x) = (4+x)(20-2x)$
$= 80 + 12x - 2x^2$

Step 4: optimize
Solving $P'(x) = 0$ yields $x = 3$

Because P is a quadratic with negative (–2) coefficient on x^2, its graph is a concave down parabola. The solution corresponds to a maximum.

price: $10 + x = \boxed{\$13}$

profit: $P(3) = \boxed{\$98}$

23. Step 1: identify variable quantities
output to optimize: revenue, R dollars
input: delay, t days
(Step 2 is skipped because the problem is not geometric.)
Step 3: construct a model (during parts a, b, and c)
a. $P(t) = 100 + 2t$ dollars per copy gives the price of the program when the release is delayed t days to package add-ons.
b. $Q(t) = 400,000 - 2300t$ copies gives the number of copies sold when the release is delayed t days to package add-ons.
c. revenue = (price)(quantity)
$\boxed{R(t) = (100 + 2t)(400,000 - 2300t) \text{ dollars gives the revenue from sales of the program when the release is delayed } t \text{ days to package add-ons.}}$
Alternate form: $R(t) = 40,000,000 + 570,000t - 4600t^2$

d. Step 4: optimize
Solving $R'(t) = 0$ yields $t \approx 61.957$ → $\boxed{62 \text{ days}}$

Because R is a quadratic with a negative (–4,600) coefficient on t^2, its graph is a concave down parabola. The solution corresponds to a maximum.
$R(62) = \boxed{\$57,657,600}$

25. Step 1: identify variable quantities
output to be optimized: cost to order and store games, C dollars
input: games needed in each order, x games per order
(Step 2 is skipped because the problem is not geometric.)
Step 3: construct a model (during parts a, b, and c)
a. i. ordering cost for x games: $20 + 4x$ dollars
   ii. number of orders annually: $\dfrac{500}{x}$ orders
   annual cost associated with ordering = (cost per order)(number of orders)
   iii.
   $$C_a = (20 + 4x) \cdot \left(\dfrac{500}{x}\right) \text{ dollars}$$

b. average number of games being stored: $\dfrac{x}{2}$ games annual cost associated with storage = (cost per game)(number of games being stored)

$$C_s = (\$6 \text{ per game}) \cdot \left(\frac{x}{2} \text{ games}\right)$$

$$\boxed{C_s = 3x \text{ dollars}}$$

c. cost to order and store games = (cost to order) + (cost to store)

$$C = C_a + C_s$$

$$= (20 + 4x)\left(\frac{500}{x}\right) + 3x$$

$$\boxed{C(x) = (20 + 4x)\frac{500}{x} + 3x \text{ dollars gives the combined ordering and storage costs when } x \text{ games are ordered at a time.}}$$

d. Step 4: optimize

Solving $C'(x) = 0$ yields $x \approx 57.735$

A graph of C illustrates that C has a relative minimum at $x \approx 57.735$ and that 58 gives the minimum of the output values corresponding to whole numbers.

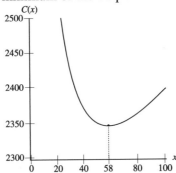

orders needed annually: $\dfrac{500}{57.735} \approx 8.660$

Because a small order (0.621) has the same fixed cost ($20) as a large order, it is necessary to compare the order and storage costs of 8 orders with that of 9 orders:
(The activity does not indicate that storage space can be prorated, so we assume that storage space must be constant throughout the year regardless of the different run sizes.)
8 orders (7 orders of 63 games, 1 order of 59)
    cost to order and store
        = (cost of first 7 orders) + (cost of final order) + (cost to store 32 games)
        $= 7 \cdot (20 + 4 \cdot 63) + (20 + 4 \cdot 59) + 6 \cdot 32$
        $= \$2352$
9 orders (8 orders of 56 games, 1 order of 52)
    cost to order and store
        = (cost of first 8 orders) + (cost of final order) + (cost to store 28 games)
        $= 8 \cdot (20 + 4 \cdot 56) + (20 + 4 \cdot 52) + 6 \cdot 28$
        $= \$2348$

Combined ordering and storage costs are minimized by $\boxed{9 \text{ orders}}$ (8 orders of 56 games, 1 order of 52).

27. a. i.   3.770 inches
        ii.   5277 bytes
        iii. 145 tracks
        iv.   765,165 bytes
        v.   1,530,330 bytes
    b. i.   6.283 inches
        ii.   8796 bytes
        iii. 91 tracks
        iv.   800,436 bytes
        v.   1,600,872 bytes
    c. i. $2\pi r$ inches
        ii. $2800\pi r$ bytes
        iii. $135(1.68-r)$ tracks
        iv. $378,000\pi r(1.68-r)$ bytes
        v. $756,000\pi r(1.68-r)$ bytes
              Solving $756,000\pi r(1.68-r)$ yields $r = 0.84$
              1,675,831 bytes

Section 4.7 Related Rates (pages 309–311)

1. $$f = 3x$$
$$\frac{d}{dt}f = \frac{d}{dt}3x$$
$$\frac{df}{dt} = 3 \cdot \frac{dx}{dt}$$
$$\boxed{\frac{df}{dt} = 3\frac{dx}{dt}}$$

3. $$k = 6x^2 + 7$$
$$\frac{d}{dy}(k) = \frac{d}{dy}6x^2 + 7$$
$$\frac{dk}{dy} = 12x + 0 \cdot \frac{dx}{dy}$$
$$\boxed{\frac{dk}{dy} = 12x\frac{dx}{dy}}$$

5. $g = e^{3x}$

$$\frac{d}{dt}[g] = \frac{d}{dt}[e^{3x}]$$

$$\frac{dg}{dt} = 3e^{3x} \cdot \frac{dx}{dt} \quad \text{by the Exponential } e^{kx} \text{ Rule}$$

$$\boxed{\frac{dg}{dt} = 3e^{3x} \frac{dx}{dt}}$$

7. $f = 62(1.02^x)$

$$\frac{d}{dt}[f] = \frac{d}{dt}[62(1.02^x)]$$

$$\frac{df}{dt} = 62\ln 1.02 (1.02^x) \cdot \frac{dx}{dt} \quad \text{by the Exponential } b^x \text{ Rule}$$

$$\boxed{\frac{df}{dt} = 62\ln 1.02 (1.02^x) \frac{dx}{dt}}$$

9. $h = 6a \ln a$

$$\frac{d}{dy}[h] = \frac{d}{dy}[6a \cdot \ln a]$$

$$\frac{dh}{dy} = \frac{d}{dy}[6a] \cdot \ln a + 6a \cdot \frac{d}{dy}[\ln a] \quad \text{by the Product Rule}$$

$$\frac{dh}{dy} = \left(6 \cdot \frac{da}{dy}\right) \cdot \ln a + 6a \cdot \left(\frac{1}{a} \cdot \frac{da}{dy}\right)$$

$$\frac{dh}{dy} = 6\ln a \cdot \frac{da}{dy} + \frac{6\cancel{a}}{\cancel{a}} \cdot \frac{da}{dy}$$

$$\boxed{\frac{dh}{dy} = 6\ln a \frac{da}{dy} + 6\frac{da}{dy}} \qquad \text{Alternate form: } \frac{dh}{dy} = (6\ln a + 6)\frac{da}{dy}$$

11. r is constant with respect to t

$$s = \pi r \sqrt{r^2 + h^2}$$

$$\frac{d}{dt}[s] = \frac{d}{dt}\left[\pi r \sqrt{r^2 + h^2}\right]$$

RHS: Rewrite $\pi r\sqrt{r^2+h^2}$ as $\pi r u^{0.5}$ where $u = r^2 + h^2$ and $\frac{du}{dt} = (0+2h)\cdot\frac{dh}{dt}$.

$$\begin{aligned}\frac{d}{dt}\pi r\sqrt{r^2+h^2} &= \frac{d}{dt}\pi r u^{0.5} \\ &= \frac{d}{du}\pi r u^{0.5}\cdot\frac{du}{dt} \quad \text{by the Chain Rule} \\ &= 0.5\pi r u^{-0.5}\cdot 2h\frac{dh}{dt} \\ &= \pi h r(r^2+h^2)^{-0.5}\frac{dh}{dt}\end{aligned}$$

$$\boxed{\frac{ds}{dt} = \frac{\pi h r}{\sqrt{r^2+h^2}}\frac{dh}{dt}}$$

Alternate form: $\dfrac{ds}{dt} = \dfrac{\pi h r\sqrt{r^2+h^2}}{r^2+h^2}\dfrac{dh}{dt}$

13. s is a constant with respect to t

Rewrite $s = \pi r\sqrt{r^2+h^2}$ as $s = \pi r\left(r^2+h^2\right)^{0.5}$

$$s = \pi r(r^2+h^2)^{0.5}$$

$$\frac{d}{dt}s = \frac{d}{dt}\pi r\cdot(r^2+h^2)^{0.5}$$

$$0 = \frac{d}{dt}\pi r\cdot(r^2+h^2)^{0.5} + \pi r\cdot\frac{d}{dt}(r^2+h^2)^{0.5}$$

last part of RHS: Rewrite $(r^2+h^2)^{0.5}$ as $u^{0.5}$

where $u = r^2+h^2$ and $\dfrac{du}{dt} = 2r\cdot\dfrac{dr}{dt} + 2h\cdot\dfrac{dh}{dt}$.

$$\begin{aligned}\frac{d}{dt}(r^2+h^2)^{0.5} &= \frac{d}{dt}u^{0.5} \\ &= \frac{d}{du}u^{0.5}\cdot\frac{du}{dt} \quad \text{by the Chain Rule} \\ &= 0.5u^{-0.5}\cdot\left(2r\frac{dr}{dt} + 2h\frac{dh}{dt}\right) \\ &= 0.5(r^2+h^2)^{-0.5}2r\frac{dr}{dt} + 0.5(r^2+h^2)^{-0.5}2h\frac{dh}{dt} \\ &= \frac{r}{(r^2+h^2)^{0.5}}\frac{dr}{dt} + \frac{h}{(r^2+h^2)^{0.5}}\frac{dh}{dt}\end{aligned}$$

$$0 = \pi \frac{dr}{dt} \cdot (r^2 + h^2)^{0.5} + \pi r \cdot \left(\frac{r}{(r^2 + h^2)^{0.5}} \frac{dr}{dt} + \frac{h}{(r^2 + h^2)^{0.5}} \frac{dh}{dt} \right)$$

$$0 = \pi (r^2 + h^2)^{0.5} \frac{dr}{dt} + \pi r \frac{r}{(r^2 + h^2)^{0.5}} \frac{dr}{dt} + \pi r \frac{h}{(r^2 + h^2)^{0.5}} \frac{dh}{dt}$$

$$0 = \left(\pi (r^2 + h^2)^{0.5} + \pi r \frac{r}{(r^2 + h^2)^{0.5}} \right) \frac{dr}{dt} + \pi r \frac{h}{(r^2 + h^2)^{0.5}} \frac{dh}{dt}$$

$$\frac{\pi h r}{(r^2 + h^2)^{0.5}} \frac{dh}{dt} = -\left(\pi (r^2 + h^2)^{0.5} + \pi r \frac{r}{(r^2 + h^2)^{0.5}} \right) \frac{dr}{dt}$$

$$\frac{dh}{dt} = -\frac{(r^2 + h^2)^{0.5}}{\pi h r} \left(\pi (r^2 + h^2)^{0.5} + \pi r \frac{r}{(r^2 + h^2)^{0.5}} \right) \frac{dr}{dt}$$

$$\frac{dh}{dt} = -\left(\frac{r^2 + h^2}{hr} + \frac{r}{h} \right) \frac{dr}{dt}$$

$$\frac{dh}{dt} = -\left(\frac{r^2 + h^2}{hr} + \frac{r^2}{hr} \right) \frac{dr}{dt}$$

$$\boxed{\frac{dh}{dt} = -\frac{2r^2 + h^2}{hr} \frac{dr}{dt}}$$

15. a. known: $g = 5$ feet

unknown: w gallons per day

$w = 31.54 + 12.97 \ln 5$

$\approx \boxed{52.414 \text{ gallons per day}}$

b. Step 1: identify variables

dependent variables: transpiration rate, w gallons per day

tree girth, g feet

independent variable: time, t years

Step 2: write an equation relating the dependent variables

$w = 31.54 + 12.97 \ln g$

Step 3: differentiate both sides of the equation with respect to the independent variable

$$\frac{d}{dt}(w) = \frac{d}{dt}(31.54 + 12.97 \ln g)$$

$$\frac{dw}{dt} = 0 + 12.97 \cdot \frac{1}{g} \cdot \frac{dg}{dt}$$

$$\frac{dw}{dt} = \frac{12.97}{g} \cdot \frac{dg}{dt}$$

Step 4: solve for the unknown rate

known: $g = 5$ feet

$$\frac{dg}{dt} = (2 \text{ inches per year})\left(\frac{1 \text{ foot}}{12 \text{ inches}} \right)$$

$$= \frac{2}{12} \text{ feet per year}$$

unknown: $\dfrac{dw}{dt}$ gallons/day per year

$$\dfrac{dw}{dt} = \dfrac{12.97}{5} \cdot \dfrac{2}{12}$$

$$\approx \boxed{0.432 \text{ gallons/day per year}}$$

$\boxed{\text{The amount of water transpired by an oak tree with girth 5 feet is increasing by 0.432 gallons/day per year.}}$

17. Step 1: identify variables
   dependent variables: weight, w pounds
   height, h inches
   BMI, B
   independent variable: time, t years

Step 2: write an equation relating the dependent variables

$$B = 703\dfrac{w}{h^2}$$

a. known: $w = 100$ pounds
   unknown: B (unit-less)
   h inches

$$B = 703\dfrac{100}{h^2}$$

$$\boxed{B \approx \dfrac{70{,}300}{h^2}}$$

b. Step 3: differentiate both sides of the equation with respect to the independent variable
   *w is constant with respect to t

Rewrite $B = 703\dfrac{w}{h^2}$ as $B = 703wh^{-2}$

$$\dfrac{d}{dt}(B) = \dfrac{d}{dt}\, 703wh^{-2}$$

$$\dfrac{dB}{dt} = 703w(-2h^{-3}) \cdot \dfrac{dh}{dt}$$

$$\boxed{\dfrac{dB}{dt} = -1406wh^{-3}\dfrac{dh}{dt}}$$

Alternate form: $\dfrac{dB}{dt} = -\dfrac{1406w}{h^3}\dfrac{dh}{dt}$

d. Step 4: solve for the unknown rate
   known: $w = 100$ pounds
   $h = 5$ feet 3 inches
   $= (5 \text{ feet})\left(\dfrac{12 \text{ inches}}{\text{foot}}\right) + 3 \text{ inches}$
   $= 63$ inches
   $\dfrac{dw}{dt} = 0.5$ inches per year

unknown: $\dfrac{dB}{dt}$ per year

$$\dfrac{dB}{dt} = -1406(100)(63^{-3})(0.5)$$
$$\approx \boxed{-0.281 \text{ per year}}$$

19. To avoid duplicate use of d in the derivative notation, we use f for diameter.
   a. Step 1: identify variables
      dependent variables: diameter, f feet
      height, h feet
      volume, V cubic feet
      independent variable: time, t years

      Step 2: write an equation relating the dependent variables
      $$V = 0.002198 f^{1.739925} h^{1.133187}$$

      Step 3: differentiate both sides of the equation with respect to the independent variable
      *f is constant with respect to t
      $$\dfrac{d}{dt} V = \dfrac{d}{dt} 0.002198 f^{1.739925} h^{1.133187}$$
      $$\dfrac{dV}{dt} = 0.002198 f^{1.739925} (1.133187 h^{1.133187}) \cdot \dfrac{dh}{dt}$$
      $$\boxed{\dfrac{dV}{dt} \approx 0.00249 f^{1.739925} h^{1.133187} \dfrac{dh}{dt}}$$

      Step 4: solve for the unknown
      known: $h = 32$ feet
      $$f = (10 \text{ inches}) \left(\dfrac{1 \text{ foot}}{12 \text{ inches}} \right)$$
      $$= \dfrac{10}{12} \text{ foot}$$
      $$\dfrac{dh}{dt} = 0.5 \text{ feet per year}$$

      unknown: $\dfrac{dV}{dt}$ cubic feet per year

      $$\dfrac{dV}{dt} \approx 0.00249 \left(\dfrac{10}{12} \right)^{1.739925} \cdot 32^{1.133187} \cdot 0.5$$
      $$\approx \boxed{0.0460 \text{ cubic feet per year}}$$

   b. Steps 1 and 2 are the same as in part a.
      Step 3: differentiate both sides of the equation with respect to the independent variable
      *h is constant with respect to t
      $$\dfrac{d}{dt} V = \dfrac{d}{dt} 0.002198 f^{1.739925} h^{1.133187}$$
      $$\dfrac{dV}{dt} = 0.002198 (1.739925 f^{1.739925}) h^{1.133187} \cdot \dfrac{dh}{dt}$$

$$\boxed{\frac{dV}{dt} \approx 0.00382 f^{1.739925} h^{1.133187} \frac{dh}{dt}}$$

Step 4: solve for the unknown
known: $h = 34$ feet

$$f = (12 \text{ inches})\left(\frac{1 \text{ foot}}{12 \text{ inches}}\right)$$
$$= 1 \text{ foot}$$

$$\frac{df}{dt} = (2 \text{ inches per year})\left(\frac{1 \text{ foot}}{12 \text{ inches}}\right)$$
$$= \frac{2}{12} \text{ foot per year}$$

unknown: $\frac{dV}{dt}$ cubic feet per year

$$\frac{dV}{dt} \approx 0.00382 \cdot 1^{1.739925} \cdot 34^{1.133187} \cdot \left(\frac{2}{12}\right)$$
$$\approx \boxed{0.0347 \text{ cubic feet per year}}$$

21. a. Solving $M = 48.1 L^{0.6} K^{0.4}$ for L:
$$M = 48.1 L^{0.6} K^{0.4}$$
$$L^{0.6} = \frac{M}{48.1 K^{0.4}}$$
$$L^{0.6 \cdot \frac{5}{3}} = \left(\frac{M}{48.1 K^{0.4}}\right)^{\frac{5}{3}} \qquad \text{hint: } \frac{1}{0.6} = \frac{5}{3}$$
$$\boxed{L = \left(\frac{M}{48.1 K^{0.4}}\right)^{\frac{5}{3}}}$$

b. Step 1: identify variables
dependent variables: production, M mattresses
labor, L worker hours
volume, V thousand dollars
independent variable: time, t years

Step 2: write an equation relating the dependent variables

Rewrite $L = \left(\frac{M}{48.1 K^{0.4}}\right)^{\frac{5}{3}}$ as $L = \frac{1}{48.1^{\frac{5}{3}}} M^{\frac{5}{3}} K^{-\frac{2}{3}}$

Step 3: differentiate both sides of the equation with respect to the independent variable
*M is constant with respect to t

$$\frac{d}{dt}L = \frac{d}{dt}\left(\frac{1}{48.1^{\frac{5}{3}}} M^{\frac{5}{3}} K^{-\frac{2}{3}}\right)$$

$$\frac{dL}{dt} = \frac{1}{48.1^{\frac{5}{3}}} M^{\frac{5}{3}} \left(-\frac{2}{3} K^{-\frac{5}{3}} \cdot \frac{dK}{dt}\right)$$

$$\boxed{\dfrac{dL}{dt} = -\dfrac{2}{3}\dfrac{M^{\frac{5}{3}}}{48.1K^{\frac{5}{3}}}\dfrac{dK}{dt}}$$

c. Step 4: solve for the unknown rate

known: $L = 8$ thousand worker hours

$K = 47$ thousand dollars

$M = 48.1(8^{0.6})(47^{0.4})$

≈ 781.337 mattresses

$\dfrac{dK}{dt} = 0.5$ thousand dollars per year

unknown: $\dfrac{dL}{dt}$ thousand worker hours per year

$\dfrac{dL}{dt} = -\dfrac{2}{3}\dfrac{\left[48.1(8^{0.6})(47^{0.4})\right]^{\frac{5}{3}}}{48.1(47)^{\frac{5}{3}}}\cdot 0.5$

$\approx \boxed{-0.057\text{ worker hours per year}}$

23. a. Step 1: identify variables

dependent variables: height, v feet

distance from observer, h feet

independent variable: time, t seconds

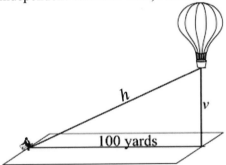

Step 2: write an equation relating the dependent variables

$h^2 = v^2 + 100^2$ by the Pythagorean Theorem

Step 3: differentiate both sides of the equation with respect to the independent variable

$\dfrac{d}{dt}\left[h^2\right] = \dfrac{d}{dt}\left[v^2 + 100^2\right]$

$2h\dfrac{dh}{dt} = 2v\dfrac{dv}{dt} + 0$

$h\dfrac{dh}{dt} = v\dfrac{dv}{dt}$

Step 4: solve for the unknown rate

known: $\dfrac{dv}{dt} = 2$ feet per second

$$v = (500 \text{ yards})\left(\frac{3 \text{ feet}}{\text{yard}}\right)$$
$$= 1500 \text{ feet}$$
$$h = \sqrt{300^2 + 1500^2}$$
$$\approx 1529.706 \text{ feet}$$

unknown: $\frac{dh}{dt}$ feet per second

Solving $\sqrt{300^2 + 1500^2} \frac{dh}{dt} = (1500)(2)$ yields $\frac{dh}{dt} \approx \boxed{1.961 \text{ feet per second}}$

b. $h \approx \boxed{1529.706 \text{ feet}}$

25. a. Step 1: identify variables
dependent variables: distance away from third base, x feet
distance away from home plate, h feet
independent variable: time, t seconds

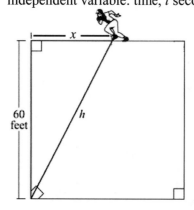

Step 2: write an equation relating the dependent variables
$h^2 = x^2 + 60^2$ by the Pythagorean Theorem

Step 3: differentiate both sides of the equation with respect to the independent variable
$$\frac{d}{dt} h^2 = \frac{d}{dt} x^2 + 60^2$$
$$2h \frac{dh}{dt} = 2x \frac{dx}{dt} + 0$$
$$h \frac{dh}{dt} = x \frac{dx}{dt}$$

Step 4: solve for the unknown rate

known: $\frac{dx}{dt} = -22$ feet per second   (negative because distance is decreasing)

$x = 30$ feet
$$h = \sqrt{60^2 + 30^2}$$
$$\approx 67.082 \text{ feet}$$

unknown: $\frac{dh}{dt}$ feet per second

Solving $\sqrt{60^2 + 30^2} \cdot \dfrac{dh}{dt} = 30(-22)$ yields $\dfrac{dh}{dt} \approx \boxed{-9.839 \text{ feet per second}}$

b. $h \approx \boxed{67.082 \text{ feet}}$

27. Step 1: identify variables
    dependent variables: volume, V cubic centimeters
    radius, r centimeters
    independent variable: time, t minutes
    Step 2: write an equation relating dependent variables
    $$V = \dfrac{4}{3}\pi r^3 \quad \text{(volume of a sphere of radius } r\text{)}$$

a. known: $r = 10$ centimeters
    unknown: V cubic centimeters
    $$V = \dfrac{4}{3}\pi(10^3)$$
    $\approx \boxed{4188.790 \text{ cubic centimeters}}$

b. $V = \dfrac{4}{3}\pi r^3$
    Step 3: differentiate both sides of the equation with respect to the independent variable
    $$\dfrac{d}{dt}V = \dfrac{d}{dt}\left(\dfrac{4}{3}\pi r^3\right)$$
    $$\dfrac{dV}{dt} = \dfrac{4}{3}\pi(3r^2)\dfrac{dr}{dt}$$
    $$\dfrac{dV}{dt} = 4\pi r^2 \dfrac{dr}{dt}$$
    Step 4: solve for the unknown rate
    known: $\dfrac{dr}{dt} = \dfrac{16-24}{30-0} = \dfrac{-4}{15}$
    ≈ -0.267 centimeters per minute
    $r = 10$ centimeters
    unknown: $\dfrac{dV}{dt}$ cubic centimeters per minute
    $$\dfrac{dV}{dt} = 4\pi\ 10\ ^2 \cdot \dfrac{-4}{15}$$
    $\approx \boxed{-335.103 \text{ cubic centimeters per minute}}$

29. Step 1: identify variables
    dependent variables: volume, V tablespoons
    height, h centimeters
    radius, r centimeters
    independent variable: time, t minutes

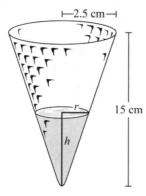

Step 2: write an equation relating dependent variables

$$V = \frac{1}{3}\pi r^2 h \qquad \text{(volume of a right circular cone with height } h \text{ and radius } r\text{)}$$

$$V = \frac{1}{3}\pi \left(\frac{h}{6}\right)^2 h \quad \text{because } r = \frac{h}{6}$$

$$V = \frac{\pi h^3}{108}$$

Step 3: differentiate both sides of the equation with respect to the independent variable

$$\frac{d}{dt}V = \frac{d}{dt}\left(\frac{\pi h^3}{108}\right)$$

$$\frac{dV}{dt} = \frac{\pi}{108}(3h^2)\frac{dh}{dt}$$

$$\frac{dV}{dt} = \frac{\pi h^2}{36}\frac{dh}{dt}$$

Step 4: solve for the unknown rate

known: $\frac{dV}{dt} = (1 \text{ tablespoon per second})\left(\frac{1 \text{ cubic centimeter}}{0.6 \text{ tablespoon}}\right)$

$\qquad = \frac{1}{0.6}$ cubic centimeter per second

$h = 6$ centimeters

unknown: $\frac{dh}{dt}$ centimeters per second

Solving $\frac{1}{0.6} = \frac{\pi(6^2)}{36}\frac{dh}{dt}$ yields $\frac{dh}{dt} \approx \boxed{5.305 \text{ centimeters per second}}$

31. Related rates involves more than one changing input variable.

Chapter 4 Review Activities (pages 313–317)

1. a. $f'(x) = -12.08x^3 + 207x^2 - 1128.98x + 1946.3$ fatalities per year gives the rate of change in the number of fatalities on charter airlines where x is the number of years after 2000, data from $3 \leq x \leq 8$.

   b. fatalities: $f(8) = 66.52 \quad \rightarrow \quad$ 67 fatalities
   ROC: $f'(8) \approx -22.500$ fatalities per year
   $f_L(x) \approx 66.52 - 22.500(x - 8)$ fatalities
   $f_L(9) \approx 44.020 \quad \rightarrow \quad \boxed{44.02 \text{ fatalities}}$

3. a. $f'(x) = -69.54x + 320.3$ fatalities per year gives the rate of change in the number of young drivers fatally injured in automobile accidents in 2007, where $x + 15$ was the age of the driver, data from $0 \leq x \leq 5$.

   b. fatalities: $f(4) = 771.88 \quad \rightarrow \quad$ 772 fatalities
   ROC: $f'(4) = 42.14$ fatalities per year (fatalities per year of age)
   $f_L(x) = \boxed{771.88 + 42.14(x - 4)}$ fatalities

   c. 20-year-old drivers: $f_L(5) = 814.02 \quad \rightarrow \quad \boxed{814 \text{ fatalities}}$
   21-year-old drivers: $f_L(6) = 856.16 \quad \rightarrow \quad \boxed{856 \text{ fatalities}}$

5. a. (The numbering on the vertical axis in the activity is off. The figure below shows the correct numbering.)

   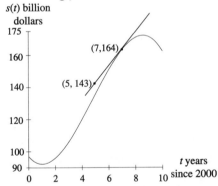

   Using points (7, 164) and (5, 143) estimated from a tangent drawn on the graph,
   slope $= \dfrac{164 - 143}{7 - 5} = \boxed{10.5 \text{ billion dollars per year}}$

   $\boxed{\text{In 1997, annual U.S. factory sales of consumer electronics were increasing by 10.5 billion dollars per year.}}$

   b. Using point (7, 164) estimated from the graph and slope 10.5 calculated in solution part a, estimate of $s(8) = 164 + 10.5 \cdot (8 - 7)$
   $= \boxed{174.5 \text{ billion dollars}}$

   c. $s(8) = \boxed{171.97 \text{ billion dollars}}$

7. $p(t) = (e^{2-t})(3^t - t^2)$

$p'(t) = \left(\dfrac{d}{dt}(e^{2-t})\right)(3^t - t^2) + (e^{2-t})\left(\dfrac{d}{dt}(3^t - t^2)\right)$ by the Product Rule

$\quad = (e^{2-t})(0-1)\ (3^t - t^2) + (e^{2-t})\ \ln 3(3^t) - 2t$

$\quad = -e^{2-t}(3^t - t^2) + (e^{2-t})\ \ln 3(3^t) - 2t$

Alternate form: $p'(t) = e^{2-t}\ (\ln 3 - 1)(3^t) - 2t + t^2$

a. Solving $p'(t) = 0$ yields to solutions:

$t \approx \boxed{0.054}$ corresponds to a relative maximum (the derivative at this point is 0)

$t \approx \boxed{1.636}$ corresponds to a relative minimum (the derivative at this point is 0)

The type of point corresponding to each solution is illustrated by a graph of p.

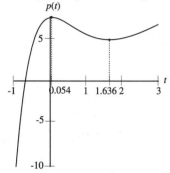

b. absolute minimum point: $-1, -13.390$ is an endpoint.

absolute maximum point: $0.054, 7.408$ is a relative maximum point.

9. a. $c(x) \approx 0.431x^3 - 15.758x^2 + 187.050x - 580.913$ chambers gives the number of non-profit national and bi-national Chambers of Commerce in the United States where x is the number of years since 1990, data from 1998 to 2007.

b. $c'(x) \approx 1.294x^2 - 31.516x + 187.050$ chambers per year gives the rate of change in the number of non-profit national and bi-national Chambers of Commerce in the United States, where x is the number of years since 1990.

c. Solving $c'(x) = 0$ yields two solutions:

$x \approx 10.250$ with corresponding output $c(10.250) \approx 145.387$

$x \approx 14.099$ with corresponding output $c(14.099) \approx 133.081$

Because c is a cubic, it is continuous and these are the only two relative extreme points.
The output values at the endpoints of the interval are

1998: $c(8) \approx 127.868$

2007: $c(17) \approx 164.582$

Comparing these four output values shows the following extreme points:

absolute minimum point: (8, 127.868)

relative maximum point: (10.250, 145.387)

relative minimum point: (14.099, 133.081)

absolute maximum point: (17, 164.582)

11. $x = -2$ and $x = 2$: relative and absolute minimum

   Both inputs correspond to the same minimum value of $y = 0$.

   The derivative does not exist because the limit of secants from the right does not equal the limit of secants from the left at this point.

   $x = 0$: relative maximum

   The derivative is 0 because the tangent line at $x = 0$ is horizontal.

   This graph does not indicate an absolute maximum because the graph does not indicate that the function is restricted to the interval shown in the figure.

13. $x = -0.75$: relative maximum

   The derivative is 0 because the tangent line at $x = -0.75$ is horizontal.

   $x = 0$: relative minimum

   The derivative is 0 because the tangent line at $x = 0$ is horizontal.

   This graph does not indicate any absolute extrema because the graph does not indicate that the function is restricted to the interval shown in the figure

15. a. $g'(x) = -0.0192x^2 + 0.384x - 0.988$ billion lunches per year

   b. $g'(10) \approx$ 0.932 million lunches per year 0.932 billion lunches per year

   c. Solving $g'(x) = 0$ yields two solutions:

   $x \approx 3.033$ with corresponding output $g(3.033) \approx 22.761$ billion lunches

   $x \approx 16.967$ with corresponding output $g(16.967) \approx 31.419$ billion lunches

   Because c is a cubic, it is continuous and these are the only two relative extreme points. The output values at the endpoints of the interval are

   $g(0) = 24.17$ million lunches

   $g(19) = 30.8124$ million lunches

   Comparing these four output values shows the following extreme points:

   relative and absolute minimum point: (3.033, 22.761)

   relative and absolute maximum point: (16.967, 31.419)

17. Figure *a* shows a graph that crosses the *x*-axis near $x = -2$. Figure *c* shows a graph with a relative maximum near $x = -2$. These points along with the behavior (direction, curvature, and end behavior) of these two graphs suggest that Figure *a* shows a slope graph of the graph in Figure *c*.

   Figure *c* shows a graph that crosses the *x*-axis near $x = 1$. This input values corresponds with the relative minimum in Figure *b*. The relative maximum from Figure *c* corresponds with the inflection point in Figure *b*. These points along with the general behavior of the graphs in Figures *b* and *c* suggest that Figure *c* shows a slope graph of the graph in Figure *b*.

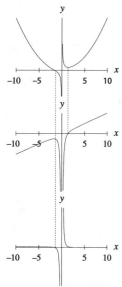

The relationships among the graphs in the three figures are as follows:

function: Figure b
derivative: Figure c
second derivative: Figure a

19. original function: $h(t) = 2t^4 - 10t^3 + 12t^2 - 5t + 6$

   first derivative: $h'(t) = 8t^3 - 30t^2 + 24t - 5$

   second derivative: $h''(t) = 24t^2 - 60t + 24$

   Solving $h''(t) = 0$ yields $t = 0.5$ and $t = 2$

21. original function: $f(u) = \dfrac{6}{u} - \dfrac{u^3}{3} + 2$

   first derivative: $f'(u) = -6u^{-2} - \dfrac{3u^2}{3}$

   Alternate form: $f'(u) = \dfrac{-6}{u^2} - u^2$

   second derivative: $f''(u) = 12u^{-3} - 2u$

   Alternate form: $f''(u) = \dfrac{12}{u^3} - 2u$

   Solving $f''(u) = 0$ yields $u \approx -1.565$ and $u \approx 1.565$

23. $D(t) = \dfrac{41.27}{1 + 40.56e^{-0.53t}}$

   $D'(t) = -41.27(1 + 40.56e^{-0.53t})^{-2}(40.56(-0.53e^{-0.53t}))$

   $\approx 887.173 e^{-0.5t}(1 + 40.56e^{-0.53t})^{-2}$

$$D''(t) \approx 887.173(-0.53e^{-0.53t})(1+40.56e^{-0.53t})^{-2}$$
$$+ 887.173e^{-0.53t}(1+40.56e^{-0.53x})^{-2}[-2(1+40.56e^{-0.53t})^{-3}(40.56(-0.53e^{-0.53t}))]$$
$$\approx -470.202e^{-0.53t}(1+40.56e^{-0.53t})^{-2} + 38142.761e^{-1.06x}(1+40.56e^{-0.53t})^{-3}$$

a. Solving $D''(t) = 0$ yields $t \approx 6.986$ → late 2007
Because D is an increasing logistic function, the solution corresponds to the point of most rapid increase.

b. $D(6.986) \approx$ 20.635%
$D'(6.986) \approx$ 5.468 percentage points per year

25. (In the activity statement, the final term of the cost function should be "5.35" instead of "5035".)

a. cost: $C(x) = 0.124x^2 + 20.7x + 5.35$ dollars gives cost when x scales are made.
marginal cost: $C'(x) = 0.248x + 20.7$ dollars per scale gives marginal cost when x scales are made.
revenue: $R(x) = 25x$ dollars gives revenue when x scales are sold.
marginal revenue: $R'(x) = 25$ dollars per scale gives marginal revenue when x scales are sold.

b. Profit is maximized when marginal cost is equal to marginal revenue.
Solving $C'(x) = R'(x)$ yields $x \approx 17.339$ → 17 scales
$P(17) = R(17) - C(17) = 31.914$ → $31.91

27. a. Step 1: identify variable quantities
output to optimize: area, A square feet
input: width, x feet
height of rectangle, y feet

Step 2: sketch

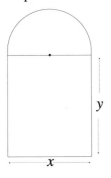

Step 3: construct a model
perimeter (semicircle + 3 sides of rectangle) = $0.5\pi x + x + 2y$
outside perimeter as stated = 15 feet
stated perimeter = semicircle + 3 sides
$$15 = 0.5\pi x + x + 2y$$
Solving $15 = 0.5\pi x + x + 2y$ for y yields $y = \dfrac{15 - (0.5\pi x + x)}{2}$
Alternate form: $y = 7.5 - (0.25\pi + 0.5)x$
area (semicircle + rectangle): $A = 0.5\pi(0.5x)^2 + xy$

$\boxed{A(x) = 0.5\pi(0.5x)^2 + x \cdot \dfrac{15-(0.5\pi x + x)}{2}}$ square feet gives the area of a Norman window $\boxed{\text{with perimeter 15 feet and width } x.}$

Alternate form: $A(x) = 0.125\pi x^2 + 7.5x - (0.25\pi + 0.5)x^2$
$= 7.5x - (0.125\pi + 0.5)x^2$

Step 4: optimize

Solving $A'(x) = 0$ yields $x \approx \boxed{4.201 \text{ feet}}$

Because A is a quadratic function with a negative coefficient $(-0.125\pi - 0.5)$ on x^2, it's graph is a concave down parabola. The relative extreme must be a maximum.

$y \approx \dfrac{15 - (0.5\pi \cdot 4.201 + 4.201)}{2} \approx \boxed{2.100 \text{ feet}}$

b. $A(4.201) \approx \boxed{15.753 \text{ square feet}}$

29. a. Step 1: identify variables
   dependent variables: length, l feet
   width, w feet
   Area, A square feet
   independent variable: time, t hours

   Step 2: write an equation relating the dependent variables
   $A = lw$

   Step 3: differentiate both sides of the equation with respect to the independent variable

   $\dfrac{d}{dt} A = \dfrac{d}{dt} lw$

   $\dfrac{dA}{dt} = \dfrac{d}{dt} l \cdot w + l \cdot \dfrac{d}{dt} w$ by the Product Rule

   $\dfrac{dA}{dt} = w \dfrac{dl}{dt} + l \dfrac{dw}{dt}$

   Step 4: solve for the unknown rate
   known: $\dfrac{dl}{dt} = 5$ feet per hour

   $\dfrac{dw}{dt} = -3$ feet per hour

   $l = 28$ feet
   $w = 22$ feet

   unknown: $\dfrac{dA}{dt}$ square feet per hour

   $\dfrac{dA}{dt} = 22 \cdot 5 + 28 \cdot (-3)$

   $= \boxed{26 \text{ square feet per hour}}$

b. Step 1: identify variables
   dependent variables: length, l feet
   width, w feet
   Perimeter, P feet
   independent variable: time, t hours

Step 2: write an equation relating the dependent variables
$$P = 2l + 2w$$
Step 3: differentiate both sides of the equation with respect to the independent variable
$$\frac{d}{dt}P = \frac{d}{dt}2l + 2w$$
$$\frac{dP}{dt} = \frac{d}{dt}2l + \frac{d}{dt}2w$$
$$\frac{dP}{dt} = 2\frac{dl}{dt} + 2\frac{dw}{dt}$$
Step 4: solve for the unknown rate

known: $\frac{dl}{dt} = 5$ feet per hour

$\frac{dw}{dt} = -3$ feet per hour

$l = 28$ feet

$w = 22$ feet

unknown: $\frac{dP}{dt}$ feet per hour

$$\frac{dP}{dt} = 2(5) + 2(-3)$$
$$= \boxed{4 \text{ feet per hour}}$$

Chapter 5 Accumulating Change: Limits of Sums and the Definite Integral

Section 5.1 An Introduction to Results of Change (pages 324–328)

1. a. The area represents the change in the amount of bacteria after t hours.
   b. i. height: thousand bacteria per hour
      width: hours
      ii. area = (height)(width)
      $$= \left(\frac{\text{thousand bacteria}}{\text{hour}}\right)(\text{hours})$$
      $$= \text{thousand bacteria}$$

3. a. The area between the input axis and the rate-of-change graph from 0 mph to 40 mph represents the stopping distance for a car traveling 40 mph. The area between the input axis and the rate-of-change graph from 0 mph to 60 mph represents the stopping distance for a car traveling 60 mph. So the area between the input axis and the rate-of-change graph from 40 mph to 60 mph represents the extra distance required to stop when a car is traveling 60 mph instead of 40 mph.
   b. i. height: feet per mph
      width: mph
      ii. area = (height)(width)
      $$= \left(\frac{\text{feet}}{\text{mph}}\right)(\text{mph})$$
      $$= \text{feet}$$

5. a. height: ppm per year
      width: year
      area: (height)(width) $= \left(\frac{\text{ppm}}{\text{year}}\right)(\text{years})$
      $$= \text{ppm}$$
   b. The region is a trapezoid with area $A = \frac{1}{2}(h_1 + h_2)w$ where
      $h_1 = 0.82$ is the height of the left side,
      $h_2 = 2$ is the height of the right side, and
      $w = 2008 - 1958 = 50$ is the width of the base.
      The area of the trapezoidal region is

$$A = \frac{1}{2}(h_1 + h_2)w$$
$$= \frac{1}{2}(0.82 + 2) \cdot 50$$
$$= \boxed{70.5 \text{ ppm}}$$

7. a. height: $\boxed{\text{miles per hour}}$
width: $\boxed{\text{hours}}$

area: (height)(width) $= \left(\dfrac{\text{miles}}{\text{hour}}\right)(\text{hours})$
$= \boxed{\text{miles}}$

b. The region between the graph of v and the input-axis is made up of three smaller regions:

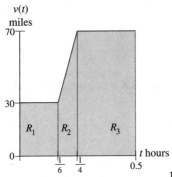

R_1 is a rectangle over $0 < t < \dfrac{1}{6}$ with

height: $h_1 = 30$

width: $w_1 = \dfrac{1}{6} - 0 = \dfrac{1}{6}$

area: $A_1 = h_1 w_1$
$= 30 \cdot \dfrac{1}{6}$
$= 5$ miles

R_2 is a trapezoid over $\dfrac{1}{6} < t < \dfrac{1}{4}$ with

height of left side: $h_1 = 30$
height of right side: $h_2 = 70$

width: $w_2 = \dfrac{1}{4} - \dfrac{1}{6} = \dfrac{1}{12}$

area: $A_2 = \dfrac{1}{2}(h_1 + h_2) \cdot w_2$
$= \dfrac{1}{2}(30 + 70) \cdot \dfrac{1}{12}$
$= \dfrac{50}{12}$ miles (or $\dfrac{25}{6} \approx 4.167$ miles)

R_3 is a rectangle over $\frac{1}{4} < t < \frac{1}{2}$ with

height: $h_2 = 70$

width: $w_3 = \frac{1}{2} - \frac{1}{4} = \frac{1}{4}$

area: $A_3 = h_2 w_3$

$= 70 \cdot \frac{1}{4}$

$= \frac{70}{4}$ miles (or $\frac{35}{2} = 17.5$ miles)

The area of the entire region is the sum of the areas of regions R_1, R_2, and R_3:

area $= A_1 + A_2 + A_3$

$= 5 + \frac{50}{12} + \frac{70}{4}$

$\approx 26.667 \quad \rightarrow \quad \boxed{26.7 \text{ miles}}$

9. Because the graph has a negative section as well as a positive section, calculations will be made as signed areas.

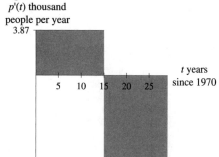

a. signed height: $h = 3.87$ thousand people per year
   width: $w = 15 - 0 = 15$ years
   signed area: $A = hw$

   $= \frac{3.87 \text{ thousand people}}{\text{year}} \cdot 15 \text{ years}$

   $= \boxed{58.05 \text{ thousand people}}$

   $\boxed{\text{From 1970 to 1985, the population of North Dakota increased by 58,050.}}$

b. signed height: $h = -7.39$ thousand people per year
   width: $w = 30 - 15 = 15$ years
   signed area: $A = hw$

   $= \frac{-7.39 \text{ thousand people}}{\text{year}} \cdot 15 \text{ years}$

   $= -110.85$ thousand people

   area: $\boxed{110.85 \text{ thousand people}}$

Area gives the magnitude of the change. In this case, the signed area is negative because the graph of p' is under the input-axis from 0 to 15. Negative signed area indicates that the change is a decrease.

From 1985 to 2000, the population of North Dakota decreased by 110,850.

c. $\begin{pmatrix} \text{change in population} \\ \text{from 1970 to 2000} \end{pmatrix} = \begin{pmatrix} \text{change in population} \\ \text{from 1970 to 1985} \end{pmatrix} + \begin{pmatrix} \text{change in population} \\ \text{from 1985 to 2000} \end{pmatrix}$

$= 58.05 - 110.85$

$= -52.80$ thousand people

The population of North Dakota was less in 2000 than it was in 1970.
There were 52.80 thousand people fewer in 2000 than in 1970.

11. (The equation in the activity has an extra t in the second piece. The equation should read
$s'(t) = \begin{cases} 0.082t - 0.39 & \text{when } 0 \leq t < 5 \\ -0.1 & \text{when } 5 \leq t \leq 7 \end{cases}$.)

Because the graph has negative sections as well as a positive section, calculations will be made as signed areas.
The graph of s' and the input-axis form three regions.

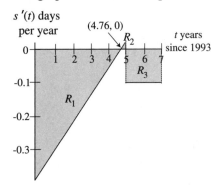

a. R_2 is a triangle over $4.76 < t < 5$ with
signed height found using the first piece of the s' equation evaluated at 5:
$h_2 = 0.082(5) - 0.39 = 0.02$ days per year
width: $w_2 = 5 - 4.76 = 0.24$ years

signed area: $A_2 = \frac{1}{2} h_2 w_2$

$= \frac{1}{2} \cdot (0.02) \cdot 0.24$

$= \boxed{0.0024 \text{ days}}$

(When a region is above the input-axis, its area is the same as its signed area.)

b. R_1 is a triangle over $0 < t < 4.76$ with
signed height: $h_1 = s'(0) = -0.39$ days per year
width: $w_1 = 4.76 - 0 = 4.76$ years

signed area: $A_1 = \dfrac{1}{2} h_1 w_1$

$\phantom{\text{signed area: } A_1} = \dfrac{1}{2} \cdot (-0.39) \cdot 4.76$

$\phantom{\text{signed area: } A_1} = -0.9282$ days

R_3 is a rectangle over $5 < t < 7$ with

signed height: $h_3 = s'(5) = -0.1$ days per year

width: $w_3 = 7 - 5 = 2$ years

signed area: $A_3 = h_3 w_3$

$\phantom{\text{signed area: } A_3} = -0.1 \cdot 2$

$\phantom{\text{signed area: } A_3} = -0.2$ days

The total signed area of regions lying below the input-axis is the sum of the signed areas of regions R_1, and R_3:

signed area $= A_1 + A_3$

$\phantom{\text{signed area }} = -0.9282 - 0.2$

$\phantom{\text{signed area }} = -1.1282$ days

The total area of the two regions is $\boxed{1.1282 \text{ days}}$.

c. $\left\{ \begin{array}{c} \text{change in stay length} \\ \text{from 1993 to 2000} \end{array} \right\} = \left(\begin{array}{c} \text{area of region} \\ \text{above the input-axis} \end{array} \right) - \left(\begin{array}{c} \text{total area of regions} \\ \text{below the input-axis} \end{array} \right)$

$\phantom{\left\{ \begin{array}{c} \text{change} \end{array} \right\}} = 0.0024 - 1.1282$

$\phantom{\left\{ \begin{array}{c} \text{change} \end{array} \right\}} = \boxed{-1.1258 \text{ days}}$

13. (Because the description includes the word "monthly," it is redundant to include "per month" in the unit of measure as was done in the answer key in the book. Also, because the context is for an "average," it is not necessary to round to whole cents in the final answer.)

Because the graph has negative sections as well as a positive section, calculations will be made as signed areas.

The graph of c' and the input-axis form three regions.

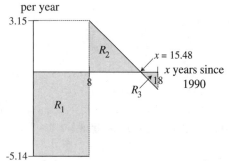

a. R_2 is a triangle over $8 < t < 15.48$ with

signed height: $h_2 = 3.15$ dollars per year

width: $w_2 = 15.48 - 8 = 7.48$ years

signed area: $A_2 = \frac{1}{2}h_2w_2$

$= \frac{1}{2} \cdot (3.15) \cdot 7.48$

$= \boxed{11.781 \text{ dollars}}$

(When a region is above the input-axis, its area is the same as its signed area.)

b. R_1 is a rectangle over $0 < t < 8$ with

signed height: $h_1 = -5.14$ dollars per year

width: $w_1 = 8 - 0 = 8$ years

signed area: $A_1 = h_1 w_1$

$= -5.14 \cdot 8$

$= -41.12$ dollars

R_3 is a triangle over $15.48 < t < 18$ with

signed height: $h_3 = -1.05$ dollars per year

width: $w_3 = 18 - 15.48 = 2.52$ years

signed area: $A_3 = \frac{1}{2}h_3w_3$

$= \frac{1}{2}(-1.05) \cdot 2.52$

$= -1.323$ dollars

The total signed area of regions lying below the input-axis is the sum of the signed areas of regions R_1, and R_3:

signed area $= A_1 + A_3$

$= -41.12 - 1.323$

$= -42.443$ dollars

The total area of the two regions is $\boxed{42.443 \text{ dollars}}$.

c. $\begin{pmatrix} \text{change in monthly bill} \\ \text{from 1990 to 2008} \end{pmatrix} = \begin{pmatrix} \text{area of region} \\ \text{above the input-axis} \end{pmatrix} - \begin{pmatrix} \text{total area of regions} \\ \text{below the input-axis} \end{pmatrix}$

$= 11.781 - 42.443$

$= \boxed{-30.662 \text{ dollars}}$

15. a. Profit is increasing when the rate-of-change graph is positive. Profit is increasing when daily production is between $\boxed{0 \text{ and } 300}$ and between $\boxed{400 \text{ and } 600}$ boxes of pencils.
    b. $\boxed{\text{NA}}$ There is not enough information given to determine the profit at any given production level.
    c. Profit is at a relative maximum when the rate-of-change graph crosses the input-axis while decreasing. Profit is at a relative minimum when the rate-of-change graph crosses the input-axis while increasing. Profit is higher than nearby profits at a production level of $\boxed{300}$ boxes, and it is lower than nearby profits at a production level of $\boxed{400}$ boxes of pencils.
    d. Profit is decreasing most rapidly when the rate-of-change graph reaches a relative minimum. The profit is decreasing most rapidly when $\boxed{350}$ boxes of pencils are produced.

e. area = (height)(width)

$$= \left(\frac{\text{dollars}}{\text{box}}\right)(\text{boxes})$$

$$= \text{dollars}$$

The units of measure of the area of the region between a graph of the rate of change of profit and the production-level axis between production levels of 100 and 200 boxes is $\boxed{\text{dollars}}$.

17. a. area = (height)(width)

$$= \left(\frac{\text{counties}}{\text{year}}\right)(\text{years})$$

$$= \boxed{\text{counties}}$$

b. The number of counties with wild turkeys in Tennessee was increasing most rapidly in mid 1978.

c. The area represents $\boxed{\text{the increase in the number of counties with wild turkeys from 1950 to mid1978}}$.

19. Answers will vary but should be similar to the following:
Area is always a positive value. Accumulated change is an interpretation of a signed area. If the area is a region between a positive portion of the rate-of-change function and the horizontal axis, the accumulated change is an increase in the function over the input interval where the rate-of-change function was positive. If the area is a region between a negative portion of the rate-of-change function and the horizontal axis, the accumulated change is a decrease in the function over the input interval where the rate-of-change function was negative.

Section 5.2   Limits of Sums and the Definite Integral
(pages 338–342)

1. a. area = (height)(width)

$$= \left(\frac{\text{thousand people}}{\text{year}}\right)(\text{years})$$

$$= \boxed{\text{thousand people}}$$

b. $\left\{\begin{array}{c}\text{units of measure} \\ \text{for } \int_{10}^{20} f(t)\,dt\end{array}\right\} = \left(\begin{array}{c}\text{units of measure} \\ \text{for } f\end{array}\right)\left(\begin{array}{c}\text{units of measure} \\ \text{for } t\end{array}\right)$

$$= \left(\frac{\text{thousand people}}{\text{year}}\right)(\text{years})$$

$$= \boxed{\text{thousand people}}$$

c. $\left\{\begin{array}{c}\text{units of measure}\\ \text{for change}\end{array}\right\} = \left(\begin{array}{c}\text{units of measure}\\ \text{for ROC}\end{array}\right)\left(\begin{array}{c}\text{units of measure}\\ \text{for interval}\end{array}\right)$

$= \left(\dfrac{\text{thousand people}}{\cancel{\text{year}}}\right)(\cancel{\text{years}})$

$= \boxed{\text{thousand people}}$

3. a. $\int_{25}^{35} g(t)dt$ represents $\boxed{\text{the change in the growth rate of blue-green algae in a river as the temperature of the water increases from 25 °C to 35 °C}}$.

   b. The area of the region between the graph of g and the t-axis from 30 °C to 40 °C represents $\boxed{\text{the change in the growth rate of blue-green algae in a river as the temperature of the water increases from 30 °C to 40 °C}}$.

5. a. 1) Calculate the width of each rectangle when four rectangles of equal width are used to cover the interval $0 \le x \le 8$:

   $$\Delta x = \dfrac{8-0}{4} = 2$$

   2) Calculate the input value for the left side of each rectangle:
   The left side of the first rectangle is the left endpoint of the interval: $x_1 = 0$
   The left sides of the remaining rectangles are calculated as
   $$x_i = x_1 + \Delta x \cdot (i-1)$$
   $$= 0 + 2(i-1)$$
   $x_1 = 0$, $x_2 = 2$, $x_3 = 4$, and $x_4 = 6$

   3) The heights of the rectangles measured at the left sides are $f(0), f(2), f(4),$ and $f(6)$, respectively. So the approximation of area is calculated as

   $$\left\{\begin{array}{c}\text{Sum of areas of}\\ \text{left rectangles}\end{array}\right\} = \sum_{i=1}^{4}(f(x_i) \cdot 2)$$
   $$= f(0) \cdot 2 + f(2) \cdot 2 + f(4) \cdot 2 + f(6) \cdot 2$$

   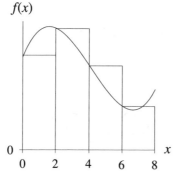

   b. 1) Calculate the width of each rectangle when four rectangles of equal width are used to cover the interval $0 \le x \le 8$:

   $$\Delta x = \dfrac{8-0}{4} = 2$$

2) Calculate the input value for the right side of each rectangle:
   The right side of the first rectangle is the left endpoint of the interval plus Δx:
   $$x_1 = 0 + 2 = 2$$
   The right sides of the remaining rectangles are calculated as
   $$x_i = x_1 + \Delta x \cdot (i-1)$$
   $$= 2 + 2(i-1)$$
   $x_1 = 2,\ x_2 = 4,\ x_3 = 6,\ \text{and}\ x_4 = 8$

3) The heights of the rectangles measured at the right sides are $f(2), f(4), f(6),$ and $f(8)$, respectively. So the approximation of area is calculated as

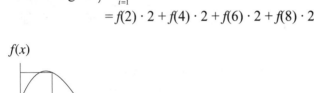

$$= f(2) \cdot 2 + f(4) \cdot 2 + f(6) \cdot 2 + f(8) \cdot 2$$

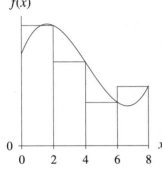

c. 1) Calculate the width of each rectangle when four rectangles of equal width are used to cover the interval $0 \le x \le 8$:
   $$\Delta x = \frac{8-0}{4} = 2$$

2) Calculate the input value for the midpoint of each rectangle:
   The midpoint of the first rectangle is the left endpoint of the interval plus half of Δx:
   $$x_1 = 0 + \frac{2}{2} = 1$$
   The midpoints of the remaining rectangles are calculated as
   $$x_i = x_1 + \Delta x \cdot (i-1)$$
   $$= 1 + 2(i-1)$$
   $x_1 = 1,\ x_2 = 3,\ x_3 = 5,\ \text{and}\ x_4 = 7$

3) The heights of the rectangles measured at the midpoints are $f(1), f(3), f(5),$ and $f(7)$, respectively. So the approximation of area is calculated as

$$\left.\begin{array}{l}\text{Sum of areas of}\\\text{right rectangles}\end{array}\right\} = \sum_{i=1}^{4}(f(x_i)\cdot 2)$$
$$= f(1)\cdot 2 + f(3)\cdot 2 + f(5)\cdot 2 + f(7)\cdot 2$$

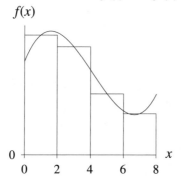

7. a. Four rectangles of equal width covering $-1 < x < 1$ have width
$$\Delta x = \frac{1-(-1)}{4} = 0.5$$

   i. The left side of the first rectangle is at the left endpoint of the interval: $x_1 = -1$
   The left sides for the remaining rectangles are calculated as
   $$x_i = x_1 + \Delta x \cdot (i-1)$$
   $$= 0.5 + 1(i-1)$$
   $x_1 = -1$, $x_2 = -0.5$, $x_3 = 0$, and $x_4 = 0.5$

   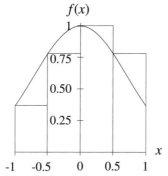

   The four left-rectangle approximation of the area is
   $$\text{area} \approx f(-1)\cdot 0.5 + f(-0.5)\cdot 0.5 + f(0)\cdot 0.5 + f(0.5)\cdot 0.5$$
   $$\approx 1.462740504 \quad \rightarrow \quad \boxed{1.463}$$

   ii. The right side of the first rectangle is
   $$x_1 = -1 + \Delta x = -1 + 0.5 = -0.5$$
   The right sides for the remaining rectangles are calculated as
   $$x_i = x_1 + \Delta x \cdot (i-1)$$
   $$= -0.5 + 0.5(i-1)$$
   $x_1 = -0.5$, $x_2 = 0$, $x_3 = 0.5$, and $x_4 = 1$

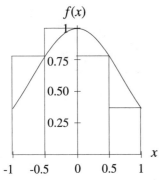

The four right-rectangle approximation of the area is
$$\text{area} \approx f(-0.5) \cdot 0.5 + f(0) \cdot 0.5 + f(0.5) \cdot 0.5 + f(1) \cdot 0.5$$
$$\approx 1.462740504 \quad \rightarrow \quad \boxed{1.463}$$

iii. The midpoint of the first rectangle is
$$x_1 = -1 + \frac{\Delta x}{2} = -1 + \frac{0.5}{2} = -0.75$$
The midpoints for the remaining rectangles are calculated as
$$x_i = x_1 + \Delta x \cdot (i-1)$$
$$= -0.75 + 0.5(i-1)$$
$x_1 = -0.75$, $x_2 = -0.25$, $x_3 = 0.25$, and $x_4 = 0.75$

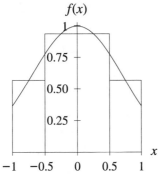

The four midpoint-rectangle approximation of the area is
$$\text{area} \approx f(-0.75) \cdot 0.5 + f(-0.25) \cdot 0.5 + f(0.25) \cdot 0.5 + f(0.75) \cdot 0.5$$
$$\approx 1.509195888 \quad \rightarrow \quad \boxed{1.509}$$

b. The area approximations found using left-rectangles and using right-rectangles are the same. (This combination of function and input interval is a special case where the function is symmetric over the interval. Dividing the interval into an even number of subintervals results in the left-rectangle and right-rectangle approximations being equal.)
These approximation underestimate the actual area by
$$1.493648266 - 1.462740504 = 0.0309077623$$
The midpoint-rectangle area approximation overestimates the actual area by
$$1.509195888 - 1.493648266 = 0.0155476122$$
The $\boxed{\text{midpoint-rectangle approximation}}$ is closest to the actual area.

9. number of rectangles: $n = 6$; interval: $0.5 \leq x \leq 2$

   width of each rectangle: $\Delta x = \dfrac{2 - 0.5}{6} = 0.25$

   a.

left side = x_i $= 0.5 + 0.25(i-1)$	height = $g(x_i)$ $= 4(\ln x)^2 \approx$	area = height · width $= g(x_i) \cdot 0.25 \approx$
0.50	1.922	0.480
0.75	0.331	0.083
1.00	0	0
1.25	0.199	0.050
1.50	0.658	0.164
1.75	1.253	0.313
	sum of areas $\approx 1.090578817 \rightarrow$	$\boxed{1.091}$

   b.

right side = x_i $= (0.5 + 0.25) + 0.25(i-1)$	height = $g(x_i)$ $= 4(\ln x)^2 \approx$	area = height · width $= g(x_i) \cdot 0.25 \approx$
0.75	0.331	0.083
1.00	0	0
1.25	0.199	0.050
1.50	0.658	0.164
1.75	1.253	0.313
2.00	1.922	0.480
	sum of areas $\approx 1.090578817 \rightarrow$	$\boxed{1.091}$

   c. midpoint of first rectangle: $0.5 + \dfrac{\Delta x}{2} = 0.5 + 0.125 = 0.625$

midpoint = x_i $= (0.5 + 0.125) + 0.25(i-1)$	height = $g(x_i)$ $= 4(\ln x)^2 \approx$	area = height · width $= g(x_i) \cdot 0.25 \approx$
0.625	0.884	0.221
0.875	0.071	0.018
1.125	0.055	0.014
1.375	0.406	0.101
1.625	0.943	0.236
1.875	1.581	0.395
	sum of areas $\approx 0.984886353 \rightarrow$	$\boxed{0.985}$

   d. The area approximations found using left-rectangles and using right-rectangles are the same. (This combination of function and input interval is a special case where the function is symmetric over the interval. Dividing the interval into an even number of subintervals results in the left-rectangle and right-rectangle approximations being equal.)
   These approximation underestimate the actual area by
   $1.090578817 - 1.019774472 = 0.070804345$
   The midpoint-rectangle area approximation overestimates the actual area by
   $1.019774472 - 0.984886353 = 0.034888119$
   The $\boxed{\text{midpoint-rectangle approximation}}$ is closest to the actual area.

11. number of rectangles: $n = 8$; interval: $0 \le t \le 24$

width of each rectangle: $\Delta t = \dfrac{24 - 0}{8} = 3$

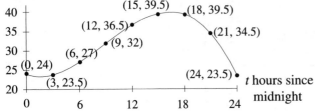

Heights are estimated from the graph.
(The answer key in the book used slightly different height estimates.)

a.

left side = t_i $= 0 + 3(i-1)$	height = $m(t_i)$ (megawatts)	area = $m(t_i) \cdot 3$ (megawatt-hours)
0	24	72
3	23.5	70.5
6	27	81
9	32	96
12	36.5	109.5
15	39.5	118.5
18	39.5	118.5
21	34.5	103.5
	sum of areas =	**769.5 megawatt-hours**

b.

right side = t_i $= (0+3) + 3(i-1)$	height = $m(t_i)$ (megawatts)	area = $m(t_i) \cdot 3$ (megawatt-hours)
3	23.5	70.5
6	27	81
9	32	96
12	36.5	109.5
15	39.5	118.5
18	39.5	118.5
21	34.5	103.5
24	23.5	70.5
	sum of areas =	**768.0 megawatt-hours**

13. Heights are estimated from the graph.

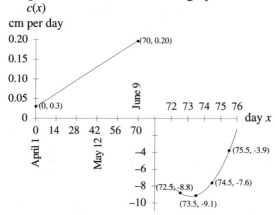

a. The region bounded by the graph of c and the t-axis from $0 \le t \le 70$ is a trapezoid.

   height of left side: $h_1 = 0.03$
   height of right side: $h_2 = 0.2$
   base width: $w = 70 - 0 = 70$

   $$\text{area} = \frac{1}{2}(h_1 + h_2)w$$
   $$= \frac{1}{2}(0.03 + 0.2) \cdot 70$$
   $$= \boxed{8.05 \text{ cm}}$$

   Between April 1 and June 9, the depth of the snow increased by an amount equivalent to 8.05 cm of water.

b. number of rectangles: $n = 4$; interval: $72 \le x \le 76$

   width of each rectangle: $\Delta x = \dfrac{76 - 72}{4} = 1$

   midpoint of first rectangle: $72 + \dfrac{1}{2} = 72.5$

midpoint = x_i = $(72.5) + 1(i-1)$	signed height = $c(x_i)$ (cm per day)	signed area = $c(x_i) \cdot 1$ (cm)
72.5	−8.8	−8.8
73.5	−9.1	−9.1
74.5	−7.6	−7.6
75.5	−3.9	−3.9
	sum of signed areas =	−29.4 cm
	area approximation =	29.4 cm

   Between June 11 and June 15, the depth of the snow decreased by an amount equivalent to 29.4 cm of water.

c. The rate of change was positive before June 9 and negative after June 11. Apparently, it stopped snowing sometime between June 9 and June 11, and the snow began to melt. There is not enough information given to know when it stopped snowing, how much more snow was received, or how much snow melted during June 10th and 11th.

15. a.

$n \to \infty$	$\sum_{i=1}^{n} \left(\dfrac{1.9}{18e^{-0.04t_i}} + 0.1 \right) \Delta t$
5	10.0724
10	10.0724
20	10.0724
40	10.0724
$\lim\limits_{n \to \infty} \left[\sum_{i=1}^{n} r(t_i) \Delta t \right]$	$\approx$ 10.072 million retirees

Note: Each line of the limit-of-sums table represents the result of a sum-of-areas table as shown in Activities 9 through 14.

b. Between 1965 and 1975, the number of Americans who were within one year of retirement increased by 10.072 million.

17. a. A graph of f shows that f is positive for part of the interval from 0 to 40 and negative for the remainder of the interval.
Solving $f(x) = 0$ yields $x \approx 21.578$.

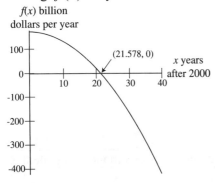

The projected total assets in the Social Security trust fund are growing from 2000 through mid-2022.

The projected total assets in the Social Security trust fund are declining from mid-2022 through 2040.

b. The amount in the trust fund will be at a maximum at the input value where the rate of change crosses the horizontal axis: mid-2022.

c. Limit of sums numerical estimation starting with 10 subintervals and incrementing by multiplying by 2:

$n \to \infty$	$\sum_{i=1}^{n} \left(-0.365 x_i^2 - 0.295 x_i + 176.32 \right) \Delta x$
10	−950.400
20	−965.000
40	−968.650
80	−969.563
160	−969.791
320	−969.848
640	−969.862
1280	−969.865
2560	−969.866
$\lim\limits_{n \to \infty} \left[\sum_{i=1}^{n} f(x_i) \Delta x \right]$	$\approx$ −968.9 billion dollars
	area $\approx$ 968.9 billion dollars

5.2 Limits of Sums and the Definite Integral

Solutions to Odd-Numbered Activities

The amount in the trust fund is projected to decline by $968.9 billion between 2000 and 2040.

d. To use the net change to project how much money will be in the trust fund at the end of 2040, it is necessary to also know how much money was in the trust fund at the end of 2000.

19. a. Limit of sums numerical estimation starting with 8 subintervals and incrementing by multiplying by 2:

$n \to \infty$	$\lim_{n \to \infty} \sum_{i=1}^{n} \left(\dfrac{13.785}{t_i} \right) \Delta t$
8	17.853
16	17.896
32	17.907
64	17.910
128	17.910
256	17.911
512	17.911
1024	17.911

$$\lim_{n \to \infty} \sum_{i=1}^{n} \left(\dfrac{13.785}{t_i} \right) \Delta t \approx \boxed{17.9 \text{ grams}}$$

b. A laboratory mouse can be expected to gain 17.9 grams in weight between weeks 3 and 11.

c. $4 + 17.9 = \boxed{21.9 \text{ grams}}$

21. a. Limit of sums numerical estimation starting with 10 subintervals and incrementing by multiplying by 2:

$n \to \infty$	$\sum_{i=1}^{n} \left(3.94 t_i^{3.55} e^{-1.35 t_i} \right) \Delta t$
10	12.506
20	12.524
80	12.525
160	12.525
320	12.525

$$\lim_{n \to \infty} \sum_{i=1}^{n} \left(3.94 t_i^{3.55} e^{-1.35 t_i} \right) \Delta t \approx \boxed{12.5 \text{ thousand barrels}}$$

b. $\int_0^{10} r(t)\, dt \approx 12.5$ thousand barrels

23. a. $F'(x) \approx 1.921(1.00019^x)$ dollars per day gives the rate of change of the amount in a continuously compounded interest bearing account, data from 365 to 3,285 days.

b. Limit of sums numerical estimation starting with 8 subintervals and incrementing by multiplying by 2:

$n \to \infty$	$\sum_{i=1}^{n} F'(x)\Delta x$
10	10,138.5108
20	10,140.0514
40	10,140.4366
80	10,140.5329
160	10,140.5569
320	10,140.5630
640	10,140.5645
1280	10,140.5648
2560	10,140.5649

$$\lim_{n \to \infty} \sum_{i=1}^{n} F'(x)\Delta x \approx \boxed{\$10,140.56}$$

c. $\int_0^{3650} F'(x)dx \approx \$10,140.56$

d. the initial value of the investment

25. Answers will vary but should include the following points:
    The accumulated change over an interval and the definite integral with the same limiting values are equivalent when considering the same continuous function. These values may be positive, negative, or zero. The area over an interval on which the continuous function being considered has negative values will differ from the accumulated change and the definite integral. The area will not ever have a negative value.

Section 5.3 Accumulation Functions (pages 350–353)

Estimates for partial grid boxes will vary but should be close to those given in the solutions presented here.

1. a. box height: 1 thousand dollars per week
   box width: 4 weeks
   box area: $\boxed{\$4000}$

   b. the profit (loss) for a new business during its first t weeks of operation

   c.

x	Additional Boxes	Additional Signed Area $\int_{x-4}^{x} p(t)dt$	Acc. Change $\int_0^x p(t)dt$	x	Additional Boxes	Additional Signed Area $\int_{x-4}^{x} p(t)dt$	Acc. Area $\int_0^x p(t)dt$
0			0	28	0.75 over	3	52
4	1 under 0.25 over	−4 + 1	−3	32	0.5 under	−2	50
8	1.75 over	7	4	36	1.5 under	−6	44
12	3 over	12	16	40	2.25 under	−9	35
16	3.25 over	13	29	44	2.25 under	−9	26
20	3 over	12	41	48	1.5 under	−6	20
24	1.75 over	7	49	52	0.25 under 0.75 over	−1 + 3	22

d.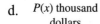
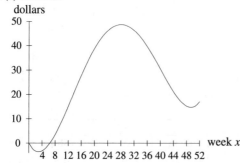

3. a. box area: $(0.5 \text{ mm per day})(2 \text{ days}) = 1 \text{ mm}$

x	Additional Boxes	Additional Signed Area $\int_{x-1}^{x} g(t)dt$	Acc. Change $\int_{0}^{x} g(t)dt$	x	Additional Boxes	Additional Signed Area $\int_{x-1}^{x} g(t)dt$	Acc. Area $\int_{0}^{x} g(t)dt$
1			0	15	2 over	2	32.75
3	8.5 over	8.5	8.5	17	1.5 over	1.5	34.25
5	7 over	7	15.5	19	1.25 over	1.25	35.5
7	5.5 over	5.5	21	21	0.9 over	0.9	36.4
9	4 over	4	25	23	0.7 over	0.7	37.1
11	3.25 over	3.25	28.25	25	0.5 over	0.5	37.6
13	2.5 over	2.5	30.75	27	0.3 over	0.3	37.9

b.
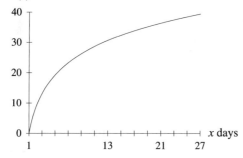

c. $G(x) = \boxed{\int_{1}^{x} g(t)dt}$

d. Over the 26 day period from the end of the first day to the end of the 27th day, the plant grew 37.9 mm.

5. a. The relative maximum point in the rate-of-change graph indicates that the number of subscribers was increasing most rapidly at that time.
   b. the change in the number of subscribers during the provider's first year

c. box area: (10 subscribers per week)(4 weeks) = 40 subscribers

t	Additional Boxes	Additional Signed Area $\int_{t-1}^{t} n(x)dx$	Acc. Change $\int_{0}^{t} n(x)dx$	t	Additional Boxes	Additional Signed Area $\int_{t-1}^{t} n(x)dx$	Acc. Area $\int_{0}^{t} n(x)dx$
0			0	28	5.5 over	220	1270
4	1.25 over	50	50	32	3.75 over	150	1420
8	2.25 over	90	140	36	2.25 over	90	1510
12	3.75 over	150	290	40	1.25 over	50	1560
16	5.5 over	220	510	44	0.6 over	24	1584
20	6.75 over	270	780	48	0.4 over	16	1600
24	6.75 over	270	1050	52	0.25 over	10	1610

d.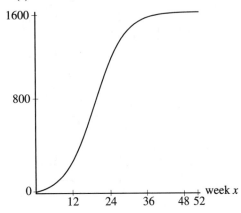

7. a.

x	0	8	18	35	47	55
additional signed area		−7.1	−8.3	30.4	25.4	−11.1
$\int_{0}^{x} r(t)dt$	0	−7.1	−15.4	15	40.4	29.3

b. During the first 18 trading days, the price of the technology stock dropped by $15.40. During the next 31 trading days (between days 18 and 47), the price of the technology stock increased by $55.80.

c. $127 + 29.3 = \boxed{\$156.3}$

d. (The table in the answer key is set vertically instead of horizontally.)

	$0 < x < 8$	$8 < x < 18$	$18 < x < 35$	$35 < x < 47$	$47 < x < 55$
Direction (magnitude)	Decreasing faster	Decreasing slower	Increasing faster	Increasing slower	Decreasing faster
Curvature	Concave down	Concave up	Concave up	Concave down	Concave down

e.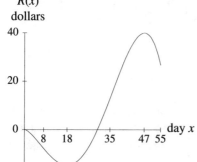

9. a. 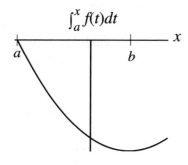   b.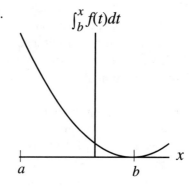

11. a. 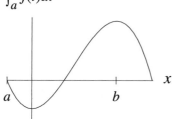   b.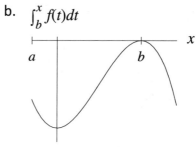

13. a. 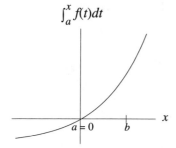   b.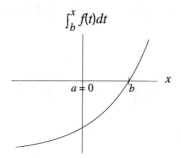

15. derivative graph: |b| — The derivative of a concave down function with a relative maximum at $x = a$ is a decreasing function that crosses the x-axis at $x = a$.
    accumulation graph: |f| — A concave down function that crosses the x-axis at $x = 0$ and $x = b$ and has a relative maximum at $x = a$ has an accumulation function graph with a relative minimum at $x = 0$, a relative maximum at $x = b$, and an inflection point at $x = a$.

17. derivative graph: |f| — The function shown in the figure has an inflection point at $x = 0$ where it crosses the x-axis while increasing and has a relative maximum at $x = c$. It's derivative function will have a relative minimum and touch but not cross the x-axis at $x = 0$ and cross the x-axis while decreasing at $x = c$.
    accumulation graph: |e| — The function shown in the figure is negative increasing and concave down to the left of $x = 0$ where it crosses the x-axis and has an inflection point, has an inflection point from concave up to concave down at $x = b$, has a relative maximum at $x = c$, and crosses the x-axis while decreasing at $x = d$. It's accumulation function graph will be positive decreasing to the left of $x = 0$, be positive increasing over $0 < x < d$ with an inflection point at $x = c$ and a relative maximum at $x = d$.

19. Assume that the output values given in the table are indicative of the behavior of the function over all real numbers within the input interval given.
    The function f is decreasing over $0 < t < 4$ so the rate-of-change function must be negative everywhere over the same interval. The function is decreasing by larger amounts as t increases, so the rate-of-changes will get larger (farther from zero). The output values for g fit this description. Because f is positive for inputs less than $t = 2$ and negative for inputs greater than $t = 2$, the accumulation function will be increasing over $0 < t < 2$ and decreasing over $2 < t < 4$. The output values for h fit this description.
    g: |rate-of-change function|
    h: |accumulation function|

21. Answers will vary but should be similar to the following:
    When a rate-of-change function is negative, the accumulation graph will be decreasing. If the rate-of-change function is negative and increasing, the accumulation graph will be decreasing but at a slower rate as the input values increase.

Section 5.4 The Fundamental Theorem (pages 363–365)

1. input: hours
    ROC output: billion KW per hour

    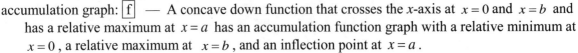

    = billion KW

5.4 The Fundamental Theorem
Solutions to Odd-Numbered Activities

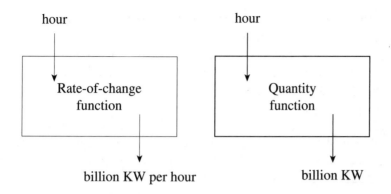

3. input: units
   ROC output: dollars per unit
   quantity function output: $\left(\begin{array}{c}\text{output units}\\\text{of ROC}\end{array}\right) \cdot \left(\begin{array}{c}\text{input units}\\\text{of ROC}\end{array}\right) = \dfrac{\text{dollars}}{\text{unit}} \cdot \text{units}$
   $= \text{dollars}$

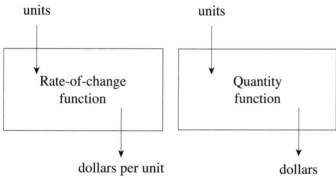

5. input: years
   ROC output: people per year
   quantity function output: $\left(\begin{array}{c}\text{output units}\\\text{of ROC}\end{array}\right) \cdot \left(\begin{array}{c}\text{input units}\\\text{of ROC}\end{array}\right) = \dfrac{\text{people}}{\text{year}} \cdot \text{years}$
   $= \text{people}$

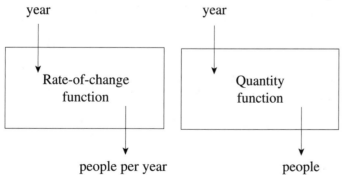

7. a. function of t
   b. function of x
   c. number

9. a. number
   b. function of x
   c. function of t

11. $\int (32x^3 + 28x - 8.5)dx = \dfrac{32x^{3+1}}{3+1} + \dfrac{28x^{1+1}}{1+1} - \dfrac{8.5x^{0+1}}{0+1} + C$

    $= \dfrac{32}{4}x^4 + \dfrac{28}{2}x^2 - 8.5x + C$

    $= \boxed{8x^4 + 14x^2 - 8.5x + C}$

13. Rewrite $\int \left(\dfrac{10}{x^6} + 3\sqrt[3]{x} + 2.5 \right) dx$ as $\int \left(10x^{-6} + 3x^{\frac{1}{3}} + 2.5 \right) dx$

    $\int \left(10x^{-6} + 3x^{\frac{1}{3}} + 2.5 \right) dx = \dfrac{10x^{-6+1}}{-6+1} + \dfrac{3x^{\frac{1}{3}+1}}{\frac{1}{3}+1} + \dfrac{2.5x^{0+1}}{0+1} + C$

    $= \dfrac{10}{-5}x^{-5} + \dfrac{3}{\frac{4}{3}}x^{\frac{4}{3}} + 2.5x + C$

    $= \boxed{-2x^{-5} + 2.25\sqrt[3]{x^4} + 2.5x + C}$

15. $S(m) = 300m^2 + 5m + C$ DVDs

17. $P(t) = \dfrac{-0.073t^4}{4} + \dfrac{1.422t^3}{3} - \dfrac{11.34t^2}{2} + 9.236 + C$ percentage points

19. $f(t) = t^2 + 2t$

    $F(t) = \dfrac{t^{2+1}}{2+1} + \dfrac{2t^{1+1}}{1+1} + C$

    $= \dfrac{t^3}{3} + t^2 + C$

    Solving $F(12) = 700$ yields $C = -20$

    $\boxed{F(t) = \dfrac{t^3}{3} + t^2 - 20}$

21. Rewrite $f(z) = \dfrac{1}{z^2} + z$ as $f(z) = z^{-2} + z$

    $F(z) = \dfrac{z^{-2+1}}{-2+1} + \dfrac{z^{1+1}}{1+1} + C$

    $= -z^{-1} + 0.5z^2 + C$

    $= \dfrac{-1}{z} + 0.5z^2 + C$

Solving $F(2) = 1$ yields $C = -0.5$

$$\boxed{F(z) = \frac{-1}{z} + 0.5z^2 - 0.5}$$

23. a. $f(t) = 0.8t - 15.9$ gallons/vehicle per year

$$F(t) = \int (0.8t - 15.9) dt$$
$$= 0.4t^2 - 15.9t + C$$

Solving $F(10) = 712$ yields $C = 831$

$$\boxed{F(t) = 0.4t^2 - 15.9t + 831 \text{ gallons/vehicle}}$$

b. The specific antiderivative in part *a* is the formula for the accumulation function of f shifted up so that it is passing through the point (10, 712).

25. a. $\int (6x^{-2} + 7) dx = -6x^{-1} + 7x + C$

b. $\dfrac{d}{dx} \int f(x) dx = \dfrac{d}{dx} \int (-6x^{-1} + 7x + C) dx$
$$= 6x^{-2} + 7$$

27. a. $\int 25x^{-4} dx = \dfrac{25x^{-3}}{-3} + C$

b. $\dfrac{d}{dx} \int f(x) dx = \dfrac{d}{dx} \left(\dfrac{25x^{-3}}{-3} + C \right)$
$$= 25x^{-4}$$

29. a. $\dfrac{d}{dx}(72x^{0.3} + 27x^{-0.3}) = 21.6x^{-0.7} - 8.1x^{-1.3}$

b. $\int \dfrac{df}{dx} dx = \int \left(21.6x^{-0.7} - 8.1x^{-1.3} \right) dx$
$$= 72x^{0.3} + 27x^{-0.3} + C$$

31. a. $\dfrac{d}{dx}(22x^{-3} - 22x^3) = -66x^{-4} - 66x^2$

b. $\int \dfrac{df}{dx} dx = \int \left(-66x^{-4} - 66x^2 \right) dx$
$$= 22x^{-3} - 22x^3 + C$$

33. a. acceleration caused by gravity: $a(t) = -32$ ft/sec^2

velocity: $v(t) = \int a(t) dt$
$$= \int -32 dt$$
$$= -32t + C$$

The initial velocity of the penny is given as 0 ft/sec.

Solving $v(0) = 0$ yields $C = 0$

$\boxed{v(t) = -32t \text{ feet per second gives the velocity of the penny } t \text{ seconds after the penny is dropped.}}$

b. position: $s(t) = \int v(t)dt$

$= \int -32t \, dt$

$= -16t^2 + K$

Initial height is given as 540 feet.
Solving $s(0) = 540$ yields $K = 540$

$\boxed{s(t) = -16t^2 + 540 \text{ feet gives the distance of the penny above ground level } t \text{ seconds after the penny is dropped.}}$

c. time until impact: Ground level is 0 feet.
Solving $s(t) = 0$ yields $t \approx \pm 5.809$ (only the positive result makes sense in context)

$t \approx 5.809 \quad \rightarrow \quad \boxed{5.8 \text{ seconds}}$

35. a. acceleration caused by gravity: $a(t) = -32 \text{ ft/sec}^2$

velocity: $v(t) = \int a(t)dt$

$= -32t + C$

It may be assumed that the cats fell from a resting position. That is, initial velocity may be assumed to be zero.
Solving $v(0) = 0$ yields $C = 0 \quad \rightarrow \quad v(t) = -32t$ ft/sec after t seconds.

position: $s(t) = \int v(t)dt$

$= -16t^2 + K$

The initial height is given as 66 feet.
Solving $s(0) = 66$ yields $K = 66 \quad \rightarrow \quad s(t) = -16t^2 + 66$ feet after t seconds.

time until impact: Solving $s(t) = 0$ gives $t \approx 2.031$ sec

impact velocity: $v(2.031) \approx -64.99 \quad \rightarrow \quad \approx \boxed{65 \text{ ft/sec}}$

$= \left(\dfrac{-64.99 \text{ ft}}{1 \text{ sec}}\right)\left(\dfrac{3600 \text{ sec}}{1 \text{ hour}}\right)\left(\dfrac{1 \text{ mile}}{5280 \text{ ft}}\right)$

≈ -44.31

$\rightarrow \quad \approx \boxed{44.3 \text{ mph}}$

b. Answers may vary but might be similar to the following:
The impact velocity of a falling cat is affected by air resistance. The mathematical formulas derived in part *a* do not take air resistance into account.

37. a. $f'(t) \approx -30.740t^2 + 416.225t - 168.964$ donors per year gives the rate of change of the number of donors to an athletics support organization t years after 1985, data from $0 \le t \le 15$.

b. $f(t) = \int f'(t)dt$

$\approx \int \left(-30.740t^2 + 416.225t - 168.964\right)dt$

$\approx -10.247t^3 + 208.113t^2 - 168.964t + C$

Solving $f(5) = 10,706$ yields $C \approx 7628.846$

$\boxed{f(t) \approx -10.247t^3 + 208.113t^2 - 168.964t + 7628.846 \text{ donors gives the number of donors to an athletics support organization } t \text{ years after 1985, data from } 0 \leq t \leq 15.}$

c. $f(17) \approx \boxed{14,559 \text{ donors}}$

Section 5.5   Antiderivative Formulas for Exponential, Natural Log, and Sine Functions   (pages 373–374)

1. $\int 19.4(1.07^x)dx = \boxed{19.4\left(\dfrac{1.07^x}{\ln 1.07}\right) + C}$ by the Exponential Rule

   Alternate form: $\int 19.4(1.07^x)dx \approx 286.733(1.07^x) + C$

3. $\int [6e^x + 4(2^x)]dx = \boxed{6e^x + 4\left(\dfrac{2^x}{\ln 2}\right) + C}$ by the Exponential (b^x and e^x) Rules

   Alternate form: $\int [6e^x + 4(2^x)]dx \approx 6e^x + 5.771(2^x) + C$

5. $\int \left(10^x + \dfrac{4}{x} + \sin x\right)dx = \boxed{\dfrac{10^x}{\ln 10} + 4\ln x - \cos x + C}$ by the Exponential, $\dfrac{1}{x}$, and Sine Rules

   Alternate form: $\int \left(10^x + \dfrac{4}{x} + \sin x\right)dx \approx 0.434(10^x) + 4\ln x - \cos x + C$

7. $\int (5.6\cos x - 3)dx = \boxed{5.6\sin x - 3x + C}$ by the Sine and Power Rules

9. (The answer key is missing parentheses enclosing $t\ln t - t$.)

   $\int (14\ln t + 9.6^t)dt = \boxed{14(t\ln t - t) + \dfrac{9.6^t}{\ln 9.6} + C}$ by the Natural Log and Exponential Rules

   Alternate form: $\int (14\ln t + 9.6^t)dt \approx 14t\ln t - 14t + 0.442(9.6^t) + C$

11. $t(x) = 200(0.93^x)$ $\dfrac{\text{DVDs}}{\text{week}}$

    $T(x) = \int t(x)dx$

    $\boxed{T(x) = \dfrac{200(0.93^x)}{\ln 0.93} + C \text{ DVDs}}$

13. The output units for c, dollars per units squared, may be written as

    dollars per unit per unit $= \dfrac{\text{dollars per unit}}{\text{unit}}$

The output units for the antiderivative C are

$$\left(\frac{\text{dollars per unit}}{\cancel{\text{unit}}}\right) \cdot (\cancel{\text{units}}) = \text{dollars per unit}$$

$c(x) = \dfrac{0.8}{x} + 0.38(0.01^x)$ $\dfrac{\text{dollars per unit}}{\text{unit}}$

$C(x) = \int c(x)dx$

$$\boxed{C(x) = 0.8\ln x + \dfrac{0.38(0.01^x)}{\ln 0.01} + C \text{ dollars per unit}}$$

Alternate form: $C(x) \approx 0.8\ln x - 0.083(0.01^x) + C$

15. $f(t) = t^2 + 2t$

$F(t) = \int f(t)dt$

$\quad = \dfrac{t^3}{3} + \dfrac{2t^2}{2} + C$

$\quad = \dfrac{t^3}{3} + t^2 + C$

Solving $F(12) = 700$ yields $C = -20$.

$$\boxed{F(t) = \dfrac{t^3}{3} + t^2 - 20}$$

17. $f(z) = \dfrac{1}{z^2} + e^z$

$\quad = z^{-2} + e^z$

$F(z) = \int f(z)dz$

$\quad = \dfrac{z^{-2+1}}{-2+1} + e^z + C$

$\quad = -z^{-1} + e^z + C$

Solving $F(2) = 1$ yields $C \approx -5.889$

$$\boxed{F(z) \approx -z^{-1} + e^z - 5.889}$$

19. a. $g(t) = \dfrac{0.57}{t}$ percentage points per year

$G(t) = \int g(t)dt$

$\quad = 0.57\ln t + C$ percent

Solving $G(10) = 4.95$ yields $C \approx 3.638$

$$\boxed{G(t) \approx 0.57\ln t + 3.638 \text{ percent}}$$

b. The specific antiderivative in part *a* is the formula for the accumulation function that passes through the point $(10, 4.95)$.

21. $F(t) = \int f(t)dt$

$= \int 0.140(1.15^t)dt$

$= \dfrac{0.140(1.15^t)}{\ln 1.15} + C$

a. 2005 to 2015: $\int_0^{10} f(t)dt \approx \$3.051$ million

2015 to 2020: $\int_{10}^{15} f(t)dt \approx \$4.098$ million

b. Solving $F(0) = 1$ yields $C \approx -0.0017$

$\boxed{F(t) \approx 1.0017(1.15^t) - 0.0017 \text{ million dollars gives the value of an investment } t \text{ years since 2005.}}$

23. a. $t(x) \approx 101.382(1.032^x)$ vehicles/hour per year gives the rate of change in vehicle traffic near a shopping center during peak hours, x years since 2000, data and projections for 2007 through 2012.

b. $T(x) = \int t(x)dx$

$\approx \int 101.382(1.032^x)dx$

$\approx \dfrac{101.382(1.032^x)}{\ln 1.032} + C$

$\approx 3267.095(1.032^x) + C$

Solving $T(7) = 3980$ yields $C \approx -93.050$

$\boxed{T(x) \approx 3267.095(1.032^x) - 93.050 \text{ vehicles per hour gives the vehicle traffic near a shopping center during peak hours } x \text{ years since 2000.}}$

c. $T(15) \approx 5147.264 \quad \rightarrow \quad \boxed{5147 \text{ vehicles}}$

Section 5.6   The Definite Integral—Algebraically (pages 381–384)

1. $\boxed{c}$ — Recovering a quantity function from its rate-of-change requires antidifferentiation. In order to calculate the quantity for a given input, instead of the change in quantity between two inputs, it is necessary to know the constant of integration.

3. $\boxed{c}$ — An accumulation function with specific information about an endpoint is a specific antiderivative.

5. $\boxed{b}$ — Even though the change between two input values in a quantity may be computed using the antiderivative of the rate-of-change function, it is not necessary to solve for a specific constant of integration.

7. $\boxed{a}$ — The slope of the line tangent to a function at a specific point is the derivative of the function at that point.

9. a. The graph of $f(x) = -4x^{-2}$ lies completely below the horizontal axis between $x = 1$ and $x = 4$.

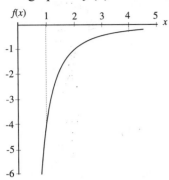

The area is calculated as

$$\text{area} = -\int_1^4 f(x)\,dx$$
$$= -(4x^{-1})\Big|_1^4$$
$$= \boxed{3}$$

b. $\int_1^4 (-4x^{-2})\,dx = \boxed{-3}$

c. When a graph of a function lies below the horizontal axis over an interval, the definite integral is the negative of the area of the region between the graph and the horizontal axis.

11. The graph of $f(x) = \dfrac{9.295}{x} - 1.472$ crosses the x-axis once in the interval $5 \le x \le 10$.

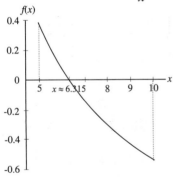

x-intercept: Solving $f(x) = 0$ yields $x \approx 6.315$.

a. Because f lies above the x-axis from $x = 5$ to $x \approx 6.315$ and below the x-axis from $x \approx 6.315$ to $x = 10$, the total area of the two regions is calculated as

$$\text{total area} \approx \int_5^{6.315} f(x)\,dx - \int_{6.315}^{10} f(x)\,dx$$
$$= (9.295 \ln x - 1.472x)\Big|_5^{6.315} - (9.295 \ln x - 1.472x)\Big|_{6.315}^{10}$$
$$\approx \boxed{1.386}$$

b. $\int_5^{10} f(x)\,dx \approx \boxed{-0.917}$

c. When a graph of a function crosses the horizontal axis within the interval of integration, the definite integral is the sum of the signed areas of the regions between the curve and the horizontal axis rather than the sum of the areas which are all positive.

5.6 The Definite Integral—Algebraically
Solutions to Odd-Numbered Activities

13. a. $\int_0^{0.005} v(t)dt = \int_0^{0.005} (-940{,}602t^2 + 19{,}269.3t - 0.3)dt$

    $= (-313{,}534t^3 + 9634.65t^2 - 0.3t)\Big|_0^{0.005}$

    $\approx 0.200 \rightarrow \boxed{0.2 \text{ miles}}$

    b. $(0.005 \text{ hours}) \cdot \left(\dfrac{60 \text{ minutes}}{\text{hour}}\right) \cdot \left(\dfrac{60 \text{ seconds}}{\text{minute}}\right) = 18$ seconds

    During the first 18 seconds of takeoff, the airplane has traveled 0.2 miles.

15. a. $\int_3^9 w(t)dt = \int_3^9 \dfrac{13.785}{t} dt$

    $= 13.785 \ln t \Big|_3^9$

    $\approx \boxed{15.144 \text{ grams}}$

    b. A mouse gains 15.144 grams between the end of the third and ninth weeks.

17. a. $\int_0^{20} r(x)dx = \int_0^{20} 1.708(0.845^x)dx$

    $= \left(\dfrac{1.708(0.845^x)}{\ln 0.845}\right)\Big|_0^{20}$

    $\approx \boxed{9.792 \ \mu\text{g/mL}}$

    The concentration of a drug increased by 9.792 μg/mL during the first 20 days after the administration was begun.

    b. $\int_{20}^{29} r(x)dx = \int_{20}^{29} (0.11875x - 3.5854)dx$

    $= (0.059375x^2 - 3.5854x)\Big|_{20}^{29}$

    $\approx \boxed{-6.084 \ \mu\text{g/mL}}$

    The concentration of a drug decreased by 6.084 μg/mL between the end of the 20th day and the end of the 29th day after the drug was administered.

    c. $\int_0^{29} r(x)dx = \int_0^{20} r(x)dx + \int_{20}^{29} r(x)dx$

    $\approx \boxed{3.708 \ \mu\text{g/mL}}$

    At the end of the 29th day after the drug was first administered, the concentration of the drug was 3.708 μg/mL.

19. a. $\int_0^{1.5} T(h)dh = \int_0^{1.5} (9.48h^3 - 15.49h^2 + 17.38h - 9.87)dh$

    $\approx (2.37h^4 - 5.163h^3 + 8.69h^2 - 9.87h)\Big|_0^{1.5}$

    $\approx \boxed{-0.681°F}$

    b. During the first hour and a half after the beginning of a thunderstorm, the temperature dropped 0.681 °F.

21. a. $\int_0^{35} a(t)dt$

b. $\int_0^{35} a(t)dt = \int_0^{35} (0.024t^2 - 1.72t + 22.58)dt$
$= (0.008t^3 - 0.86t^2 + 22.58t)\Big|_0^{35} \approx \boxed{79.8 \text{ mph}}$

23. a. $R'(x) \approx -0.016x^2 + 3.323x - 67.714$ thousand dollars per hundred dollars gives the approximate increase in revenue that occurs when an additional \$100 is spent on advertising when x hundred dollars is already spent on advertising.

$R(x) = \int R'(x)dx$

b. $= \int (-0.016x^2 + 3.323x - 67.714)dx$
$\approx -0.005x^3 + 1.661x^2 - 67.714x + C$
Solving $R(50) = 877$ yields $C \approx 761.524$.

$\boxed{R(x) \approx -0.005x^3 + 1.661x^2 - 67.714x + 761.524 \text{ thousand dollars gives the revenue when } x \text{ hundred dollars is spent on advertising.}}$

c. $\int_{80}^{130} R'(x)dx = R(x)\Big|_{80}^{130}$
$\approx \$5265.190 \text{ thousand}$
$\boxed{\text{Sales revenue will increase by approximately 5265 thousand dollars.}}$

25. a. $T(x) = \int T'(x)dx$
$= \int 11.4\cos(0.524x - 2.27)dx$
$= \dfrac{11.4\cos(0.524x - 2.27)}{0.524} + C$
Solving $T(7) = 73.5$ yields $x \approx 52.068$

$\boxed{T(x) = 11.4\dfrac{\sin(0.524x - 2.27)}{0.524} + 52.068 \text{ °F gives the average temperature of New York where } x = 1 \text{ in January, } x = 2 \text{ in February, and so on.}}$

b. $T(12) \approx \boxed{35.35°F}$

c. $\int_2^8 T'(x)dx \approx 40.873°F$
$\boxed{\text{The average temperature for August is 40.873 degrees warmer than the average temperature for February.}}$

27. a. $p(s) = \dfrac{40.5\sin(0.01345s - 1.5708) + 186.5}{1000}$ pulses per milliseconds

b. period $= \dfrac{2\pi}{0.01345} \approx 467.151$

$\int_0^{467.151} p(s)ds = \int_0^{467.151} (40.5\sin(0.01345s - 1.5708) + 186.5)/1000\, ds$
$\approx \boxed{87.124 \text{ pulses}}$
$\boxed{\text{During 467 milliseconds, approximately 87 pulses occur.}}$

Section 5.7 Differences of Accumulated Change (pages 389–392)

1. a.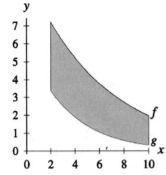

   b. $\int_{2}^{10}(f(x)-g(x))dx = \int_{2}^{10}\left(10(0.85^x)-6(0.75^x)\right)dx$

   $\approx \left(-61.631(0.85^x)+20.856(0.75^x)\right)\Big|_{2}^{10}$

   $\approx \boxed{21.785}$

3. a.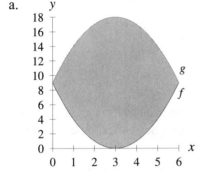

   b. $\int_{0}^{6}(g(x)-f(x))dx = \int_{0}^{6}\left((-x^2+6x+9)-(x^2-6x+9)\right)dx$

   $\approx (-0.667x^3+6x^2)\Big|_{0}^{6}$

   $= \boxed{72}$

5. a.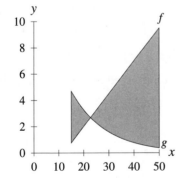

   b. Solving $f(x)=g(x)$ yields $x \approx \boxed{22.747}$

c. $\int_{15}^{50} f(x)dx - \int_{15}^{50} g(x)dx \approx \boxed{119.543}$

d. $\int_{15}^{22.747}(g(x)-f(x))dx + \int_{22.747}^{50}(f(x)-g(x))dx \approx \boxed{148.786}$

7. a.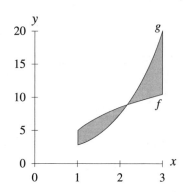

b. Solving $f(x) = g(x)$ yields $x \approx \boxed{2.188}$

c. $\int_1^3 f(x)dx - \int_1^3 g(x)dx \approx \boxed{-0.888}$

d. $\int_1^{2.188}(f(x)-g(x))dx - \int_{2.188}^3 (g(x)-f(x))dx \approx \boxed{5.622}$

9. a. Between 1998 and 2001, profit increased by $126.5 billion.

b. $\int_{1998}^{2001}(R'(x) - C'(x))dx = 126.5$

11. a. The difference in the increase in people who are sick due to the virus and the increase in people who have contracted and recovered from the virus over the first 14 days after the epidemic begins is equal to the area of region R_1.

b. The difference in the increase in people who have contracted and recovered from the virus and the increase in people who are sick due to the virus over the 14th through 20th days after the epidemic begins is equal to the area of region R_2.

c. Answers will vary but may be similar to the following:

$\int_0^{20} c(t)dt$ gives increase in people who have contracted the virus between day 0 and day 20

and $\int_0^{20} r(t)dt$ gives the increase in people who have recovered by day 20, so the difference,

$\int_0^{20}[c(t)-r(t)]dt$, will result in the increase in people who contracted the virus during the 20 day period but who have not yet recovered.

13. a. $\int_0^{11} I'(t)dt - \int_0^{11} E'(t)dt \approx \boxed{\$404.663 \text{ billion}}$

b. Because the graphs do not intersect, the answer to part *a* is the area of the region between the curves.

15. a. USPS: $p(x) \approx 0.008x^3 - 0.154x^2 + 0.606x + 2.767$ billion dollars per year gives the rate of change of revenue x years after 1990, data from 1993 to 2001.
UPS: $u(x) \approx 0.15x + 0.572$ billion dollars per year x years after 1990, data from 1993 to 2001
   b. intersection between 3 and 11: Solving $p(x) = u(x)$ yields three solutions, but only one occurs within the interval mentioned: $x \approx 8.574$
$$\int_3^{8.574} (p(x) - u(x))dx \approx \boxed{\$6.995 \text{ billion}}$$

   From 1993 to approximately midway in 1998, the revenue of USPS exceeded the revenue of UPS by approximately \$7 billion.

$$\int_{8.574}^{11} (u(x) - p(x))dx \approx \boxed{\$0.714 \text{ billion}}$$

   From midway through 1998 to the end of 2001, the revenue of UPS exceeded the revenue of USPS by approximately \$0.7 billion.

17. a. $\int_2^5 (m(t) - p(t))dt \approx 1.739$ → 1.7 hundred moths per square meter
   b. Between 1962 and 1965, there were approximately 170 moths killed by predatory beetles per square meter of canopy.

19. Answers may vary but should be similar to the following:
   Region R_1 represents the amount the profit increased by over this interval. Region R_2 represents the amount the profit decreased by over this interval. Region R_3 represents the amount the profit increased by over this interval. The integral represents the change in the profit over this interval.

Section 5.8 Average Value and Average Rate of Change (pages 399–403)

1. a. $\dfrac{\int_0^{11} s(t)dt}{11-0} \approx \boxed{67.885 \text{ billion dollars}}$

   b. $\dfrac{s(11) - s(0)}{11-0} \approx \boxed{4.948 \text{ billion dollars per year}}$

   c.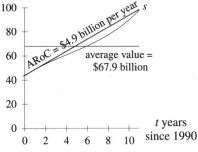

3. a. $\dfrac{\int_{89}^{110} p(t)dt}{110-89} \approx \boxed{94.730 \text{ million people}}$

   b. Solving $p(t) = 94.730$ yields $t = 99.953$ → $\boxed{\text{near the end of 2000}}$

   c. $\dfrac{p(110)-p(89)}{110-89} \approx \boxed{2.339 \text{ million people per year}}$

   d. $p(t)$ million people per year

   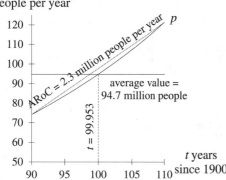

5. a. $\dfrac{\int_{6}^{20} n(x)dx}{20-6} \approx \boxed{59.668 \text{ million newspapers}}$

   b. Solving $n(x) = 59.668$ yields $x \approx 13.213$ → $\boxed{1994}$

7. a. $\dfrac{\int_{0}^{35} a(t)dt}{35-0} = \boxed{2.28 \text{ ft/sec}^2}$

   b. $v(t) = \int (0.024t^2 - 1.72t + 22.58)dt$
   $= 0.008t^3 - 0.86t^2 + 22.58t + C$
   Solving $v(0) = 0$ yields $C = 0$
   $v(t) = 0.008t^3 - 0.86t^2 + 22.58t$
   $\dfrac{\int_{0}^{35} v(t)dt}{35-0} \approx \boxed{129.733 \text{ ft/sec}}$

   c. $\int_{0}^{35} v(t)dt \approx \boxed{4540.667 \text{ feet}}$

   d. 4540.667 feet

9. a. $v(t) = -1.664t^3 + 5.597t^2 + 1.640t + 60.164$ mph gives the typical weekday speeds during the 4 P.M. to 7 P.M. rush hours on a newly widened stretch of Interstate, where t is the number of hours since 4 P.M.

   b. $\dfrac{\int_{0}^{3} v(t)dt}{3-0} \approx \boxed{68.991 \text{ mph}}$

11. a. $p(x) \approx 0.0027x^2 - 0.125x + 1.629$ dollars per minute gives the most expensive rates for a 2-minute telephone call using a long-distance carrier, where x is the number of years since 1980, based on data between 1982 and 2000.

   b. $\dfrac{\int_2^{20} p(x)dx}{20-2} \approx \$0.651$ per minute

   c. $\dfrac{p(20)-p(2)}{20-2} \approx \boxed{-\$0.065 \text{ per minute per year}}$

13. a. $C(h) = \int c(h)dh$

   $\qquad = \int \left(-0.016h^3 + 0.15h^2 - 0.54h + 2.05\right)dh$

   $\qquad = -0.004h^4 + 0.05h^3 - 0.27h^2 + 2.05h + K$

   Solving $C(0) = 3.1$ yields $K = 3.1$

   $C(h) = -0.004h^4 + 0.05h^3 - 0.27h^2 + 2.05h + 3.1$ ppm

   Solving $c(h) = 0$ yields $h \approx 7.161$ (Even though c is cubic, it has only one real root.)
   A graph of C illustrates that this solution corresponds to a relative maximum.

   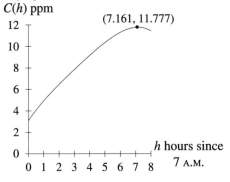

   $C(7.161) \approx 11.777$ ppm

   $\boxed{\text{The city did not exceed the 35-ppm maximum concentration.}}$

   b. $\dfrac{\int_0^8 C(h)dh}{8-0} \approx 8.663$ ppm

   $\boxed{\text{The city did not exceed the 9-ppm average concentration between 7 A.M. and 3 P.M.}}$

   c.

15. a. On the interval, $0 \le t \le 12$, $S(t)$ has a maximum output of $79.259 thousand when $t = 0.299$ (January) and $t = 6.665$ (July). $S(t)$ has a minimum output of $28.741 thousand when $t = 3.482$ (April) and $t = 9.848$ (October).

b. $\dfrac{\int_0^{12} S(t)dt}{12-0} \approx \boxed{\$52.853 \text{ thousand}}$

17. a. $v(t) \approx 1.033t + 138.413$ meters per second gives the velocity of a crack during a 60-microsecond experiment, where t microseconds gives the elapsed time since the start of the breakage.

b. $\dfrac{\int_{10}^{60} v(t)dt}{60-10} \approx \boxed{174.583 \text{ meters per second}}$

Section 5.9   Integration of Product or Composite Functions (page 409)

1. Let $u = 2x$ and $du = 2$
$$\int 2e^{2x} dx = \int e^{2x} \cdot 2 dx$$
$$= \int e^u \cdot du$$
$$= e^u + C$$
$$= \boxed{e^{2x} + C}$$

3. Let $u = 2x^2$ and $du = 4x dx$
$$\int 3xe^{2x^2} dx = \int \dfrac{3}{4} e^{2x^2} \cdot 4x dx$$
$$= \dfrac{3}{4} \int e^u \cdot du$$
$$= \dfrac{3}{4} e^u + C$$
$$= \boxed{\dfrac{3}{4} e^{2x^2} + C}$$

5. Let $u = 1 + e^x$ and $du = e^x dx$
$$\int (1+e^x)^2 e^x dx = \int u^2 du$$
$$= \dfrac{u^3}{3} + C$$
$$= \boxed{\dfrac{(1+e^x)^3}{3} + C}$$

7. Let $u = 2^x + 2$ and $du = (\ln 2)2^x\,dx$

$$\int \frac{2^x}{2^x+2}\,dx = \frac{1}{\ln 2}\int \frac{1}{2^x+2}\cdot(\ln 2)2^x\,dx$$

$$= \frac{1}{\ln 2}\int \frac{1}{u}\cdot du$$

$$= \frac{\ln u}{\ln 2} + C$$

$$= \boxed{\frac{\ln(2^x+2)}{\ln 2} + C}$$

9. By the Natural Log Rule
$$\int \ln x\,dx = \boxed{x\ln x - x + C}$$

11. Let $u = \ln x$ and $du = \frac{1}{x}dx$

$$\int \frac{\ln x}{x}\,dx = \int \ln x \cdot \frac{1}{x}\,dx$$

$$= \int u\cdot du$$

$$= \frac{u^2}{2} + C$$

$$= \boxed{\frac{(\ln x)^2}{2} + C}$$

13. Let $u = x^2 + 1$ and $du = 2x\,dx$

$$\int 2x\ln(x^2+1)\,dx = \int \ln(x^2+1)\cdot 2x\,dx$$

$$= \int \ln u\,du$$

$$= u\ln u - u + C$$

$$= \boxed{(x^2+1)\ln(x^2+1) - (x^2+1) + C}$$

15. Let $u = x^2 + 1$ and $du = 2x\,dx$

$$\int \frac{2x}{x^2+1}\,dx = \int \frac{1}{x^2+1}\cdot 2x\,dx$$

$$= \int \frac{1}{u}\cdot du$$

$$= \ln u + C$$

$$= \boxed{\ln(x^2+1) + C}$$

17. Let $u = x^3 + 5$ and $du = 3x^2 dx$

$$\int 2x^2(x^3+5)^{\frac{3}{2}}dx = \frac{2}{3}\int (x^3+5)^{\frac{3}{2}} \cdot 3x^2 dx$$

$$= \frac{2}{3}\int u^{\frac{3}{2}} \cdot du$$

$$= \frac{2}{3}\frac{u^{\frac{5}{2}}}{\frac{5}{2}} + C$$

$$= \boxed{\frac{4(x^3+5)^{\frac{5}{2}}}{15} + C}$$

19. By algebraic manipulation:

$$\int \frac{x^2+1}{x^2}dx = \int \left(1 + \frac{1}{x^2}\right)dx$$

$$= \boxed{x - \frac{1}{x} + C}$$

21. a. Let $u = 2x$ and $du = 2dx$

$$\int 2e^{2x}dx = \int e^{2x} \cdot 2dx$$

$$= \int e^u \cdot 2dx$$

$$= e^u + C$$

$$= \boxed{e^{2x} + C}$$

b. $\int_0^4 2e^{2x}dx \approx \boxed{2980}$

23. a. Let $u = \sin x$ and $du = \cos x \, dx$

$$\int (\sin x)^2 \cos x \, dx = \int (\sin x)^2 \cdot \cos x \, dx$$

$$= \int u^2 du$$

$$= \frac{u^3}{3} + C$$

$$= \boxed{\frac{(\sin x)^3}{3} + C}$$

b. $\int_0^5 (\sin x)^2 \cos x \, dx \approx \boxed{-0.294}$

Chapter 5 Review Activities (pages 411–416)

1. a. area = (height)(width)

    $$= \left(\frac{\text{billion prescriptions}}{\text{year}}\right)(\text{years})$$

    $$= \boxed{\text{billion prescriptions}}$$

    b. height: $\boxed{\text{billion prescriptions per year}}$
        width: $\boxed{\text{years}}$
    c. accumulated change: $\boxed{\text{billion prescriptions}}$

3. (The vertical axis label in the activity should be "$R(x)$ dollars per thousand miles" instead of "…per year".)
    Because the graph has a negative section as well as a positive section, calculations will be made as signed areas.
    The graph of R and the input-axis form two regions.

    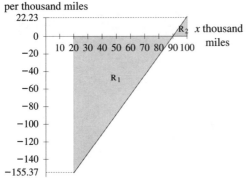

    a. R_1 is a triangle over $20 < x < 90$ with

        signed height: $h_1 = -155.37$ dollars per thousand miles
        width: $w_1 = 90 - 20 = 70$ thousand miles

        signed area: $A_1 = \frac{1}{2} h_1 w_1$

        $$= \frac{1}{2} \cdot (-155.37) \cdot 70$$

        $$= -\$5437.95$$

        area: $\boxed{\$5437.95}$

        $\boxed{\text{As the number of miles on the 2008 Ford Explorer increased from 20,000 to 90,000, the private-party resale value decreased by \$5437.95.}}$

    b. R_2 is a triangle over $90 < x < 100$

        signed height: $h_2 = 22.23$ dollars per thousand miles
        width: $w_2 = 100 - 90 = 10$ thousand miles

signed area: $A_2 = \frac{1}{2}h_2 w_2$

$= \frac{1}{2} \cdot (22.23) \cdot 10$

$= \boxed{\$111.15}$

As the number of miles on the 2008 Ford Explorer increases from 90,000 to 100,000, the private-party resale value increases by $111.15.

c. $111.15 - (-5437.95) = \boxed{\$5326.80}$

5. a. area = (height)(width)

$= \left(\frac{\text{percentage points}}{\cancel{\text{year}}}\right)(\cancel{\text{years}})$

$= \boxed{\text{percent}}$

b. $\left\{\begin{array}{c}\text{units of measure} \\ \text{for } \int_7^9 p(t)dt\end{array}\right\} = \left(\begin{array}{c}\text{units of measure} \\ \text{for } p\end{array}\right)\left(\begin{array}{c}\text{units of measure} \\ \text{for } t\end{array}\right)$

$= \left(\frac{\text{percentage points}}{\cancel{\text{year}}}\right)(\cancel{\text{years}})$

$= \boxed{\text{percent}}$

c. $\left\{\begin{array}{c}\text{units of measure} \\ \text{for change}\end{array}\right\} = \left(\begin{array}{c}\text{units of measure} \\ \text{for ROC}\end{array}\right)\left(\begin{array}{c}\text{units of measure} \\ \text{for interval}\end{array}\right)$

$= \left(\frac{\text{percentage points}}{\cancel{\text{year}}}\right)(\cancel{\text{years}})$

$= \boxed{\text{percent}}$

7. a. 1) Calculate the width of each rectangle when six rectangles of equal width are used to cover the interval $2 \le t \le 14$:

$\Delta t = \frac{14-2}{6} = 2$

2) Calculate the input value for the midpoint of each rectangle:
The midpoint of the first rectangle is the left endpoint of the interval plus half of Δt:

$t_1 = 2 + \frac{2}{2} = 3$

The midpoints of the remaining rectangles are calculated as

$t_i = t_1 + \Delta t \cdot (i-1)$

$= 3 + 2(i-2)$

$t_1 = 3, \ t_2 = 5, \ t_3 = 7, \ t_4 = 9, \ t_5 = 11, \ \text{and } t_6 = 13$

3) The heights of the rectangles measured at the midpoints are $p'(3), p'(5), p'(7), p'(9), p'(11),$ and $p'(13)$, respectively.

The heights of the rectangles are estimated from the graph of p'.

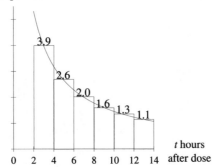

$$\begin{pmatrix}\text{Sum of areas of} \\ \text{midpoint rectangles}\end{pmatrix} = \sum_{i=1}^{6}(p'(t_i) \cdot 2)$$

$$= p'(3) \cdot 2 + p'(5) \cdot 2 + p'(7) \cdot 2 + p'(9) \cdot 2 + p'(11) \cdot 2 + p'(13) \cdot 2$$

$$\approx 3.9 \cdot 2 + 2.6 \cdot 2 + 2.0 \cdot 2 + 1.6 \cdot 2 + 1.3 \cdot 2 + 1.1 \cdot 2$$

$$= \boxed{25.0 \text{ percent}}$$

b.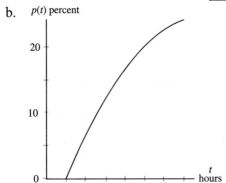

9. (In the activity statement the leading coeffiecient of the function h should be negative. That is, $h(t) = -0.0006t^3 + \ldots$

 a.

$n \to \infty$	$\lim_{n \to \infty} \sum_{i=1}^{n}(-0.0006t^3 + 0.01614t^2 - 0.1102t + 0.0163)\Delta t$
16	-1.709
32	-1.709
64	-1.709
128	-1.709

$$\lim_{n \to \infty} \sum_{i=1}^{n}(-0.0006t^3 + 0.01614t^2 - 0.1102t + 0.0163)\Delta t \approx \boxed{-1.7 \text{ days}}$$

b. A graph of h shows that the function lies entirely below the horizontal axis over $1 < t < 17$.

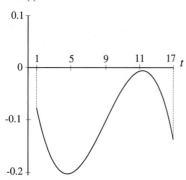

Because the graph of h lies below the t-axis over the entire interval $1 < t < 17$, the answer in part a is the negative of the area of the region.

c. Between 1990 and 2007, the average length of a stay in a U.S. community hospital decreased by approximately 1.7 days.

11. a. box height: 0.25 percentage point per year
    box width: 1 year
    box area: 0.25 percentage point

x	Additional Boxes	Additional Signed Area $\int_{x-1}^{x} s'(t)\,dt$	Acc. Change $\int_{0}^{x} s'(t)\,dt$	x	Additional Boxes	Additional Signed Area $\int_{x-1}^{x} s'(t)\,dt$	Acc. Area $\int_{0}^{x} s'(t)\,dt$
0			0	5	6.3 over	1.575	2.1875
1	0 over	0	0	6	6.8 over	1.7	3.8875
2	0.05 over	0.0125	0.0125	7	2.3 over	0.575	4.4625
3	0.4 over	0.1	0.1125	8	0.5 over	0.125	4.5875
4	2 over	0.5	0.6125				

b. $\int_{0}^{x} s'(t)\,dt$

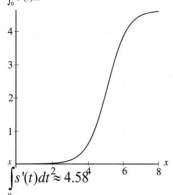

c. $\int_{0}^{x} s'(t)\,dt \approx 4.58$

Between 2001 and 2017, the percentage of unmarried men age 15 and up increased by 4.58 percentage points.

d. s(t) percent

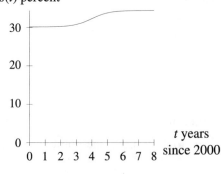

30.1 + 4.58 = 34.68 percent

13. (In the activity, both times the indefinite integral notation is used it should have a lower limit of 1 instead of 0. Also in part d, the indefinite integral should be referred to as capital P instead of lower case.)

a.
x	1	3.7	7.1	14
additional signed area		3.24	−2.48	−3.62
$\int_1^x p'(t)\,dt$	0	3.24	0.76	−2.86

b. The percentage of marriages ending in divorce during the xth year decreases by 6.1 % between the 8th month of the 4th and the end of the 14th year of marriage.

c. $4.7 - 2.86 = \boxed{1.84 \text{ percent}}$

d. One possible graph

P(t) percent

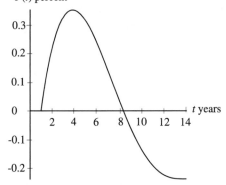

15. $r(x) = -0.0084x + 0.166$ billion prescriptions per year

$R(x) = \int (-0.0084x + 0.166)\,dx$

$= \dfrac{-0.0084x^2}{2} + 0.166x + C$

$\boxed{R(x) = -0.0042x^2 + 0.166x + C \text{ billion prescriptions}}$

17. a. $n(t) = 35.49t^2 - 320.32t + 887.88$ million dollars per year

$$N(t) = \int (35.49t^2 - 320.32t + 887.88)dt$$

$$= \frac{35.49t^3}{3} - \frac{320.32t^2}{2} + 887.88t + C$$

Solving $N(1) = 3.69$ yields $C = -735.86$.

$$\boxed{N(t) = 11.83t^3 - 160.16t^2 + 887.88t - 735.86 \text{ million dollars}}$$

b. The function N is the accumulation function of n shifted to go through the point (1, 3.69).

19. $\int \left[2.4e^x - 7(3^x) \right] = \boxed{2.4e^x - \frac{7(3^x)}{\ln 3} + C}$

21. Rewrite the first term as $\int \frac{3}{8^x} dx = \int -3(8^{-x})(-1)dx$

$$= \int -3(8^u)du \quad \text{where } u = -x \text{ and } du = -dx$$

$$= -3 \cdot \frac{8^u}{\ln 8} + c$$

$$= \frac{-3(8^{-x})}{\ln 8} + c$$

$$\int \left[\frac{3}{8^x} - 5\cos x - \ln x \right] dx = \boxed{\frac{-3(8^{-x})}{\ln 8} - 5\sin x - (x\ln x - x) + C}$$

23. $V(t) = \int 1.83(1.0546^t)dt$

$$= 1.83 \cdot \frac{1.0546^t}{\ln 1.0546} + C$$

Solving $V(8) = 51.811$ yields $C \approx -0.858$

$$\boxed{V(t) \approx \frac{1.83(1.0546^t)}{\ln 1.0546} - 0.858 \text{ billion dollars}}$$

25. The graph of $h(x) = 1.7^x - \frac{2.83}{x}$ crosses the x-axis once in the interval $0.1 \le x \le 4$.

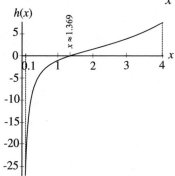

x-intercept: Solving $h(x) = 0$ yields $x \approx 1.369$.

a. Because h lies below the *x*-axis from $x = 0.1$ to $x \approx 1.369$ and above the *x*-axis from $x \approx 1.369$ to $x = 4$, the total area of the two regions is calculated as

$$\text{total area} \approx -\int_{0.1}^{1.369} h(x)dx + \int_{1.369}^{4} h(x)dx$$

$$= -\left(\frac{1.7^x}{\ln 1.7} - 2.83\ln x\right)\Bigg|_{0.1}^{1.369} + \left(\frac{1.7^x}{\ln 1.7} - 2.83\ln x\right)\Bigg|_{1.369}^{4}$$

$$\approx \boxed{14.305}$$

b. $\int_{0.1}^{4} h(x)dx \approx \boxed{3.313}$

c. When a graph of a function crosses the horizontal axis within the interval of integration, the definite integral is the sum of the signed areas of the regions between the curve and the horizontal axis rather than the sum of the areas which are all positive.

27. $p(t) = \begin{cases} -0.056t - 0.031 & \text{for } 0 \leq t \leq 5 \\ -0.12t + 0.79 & \text{for } 5 < t \leq 8 \end{cases}$

$\int p(t)dt = \begin{cases} -0.028t^2 - 0.031t + c_1 & \text{for } 0 \leq t \leq 5 \\ -0.06t^2 + 0.79t + c_2 & \text{for } 5 < t \leq 8 \end{cases}$

a. $\int_0^5 p(t)dt = (-0.028t^2 - 0.031t)\Big|_0^5$

$= \boxed{-0.855 \text{ million barrels}}$

Between 2000 and 2005, daily U.S. petroleum production decreased by 0.855 million barrels.

b. $\int_5^8 p(t)dt = (-0.06t^2 - 0.79t)\Big|_5^8$

$= \boxed{0.03 \text{ million barrels}}$

Between 2005 and 2008, daily U.S. petroleum production increased by 0.03 million barrels.

c. $\int_0^8 p(t)dt = \int_0^5 p(t)dt + \int_5^8 p(t)dt$

$= \boxed{-0.825 \text{ million barrels}}$

Between 2000 and 2008, daily U.S. petroleum production decreased by 0.825 million barrels.

29. a. Solving $f(x) = g(x)$ yields two solutions

$x \approx \boxed{1.458}$

$x \approx 4.894$ which is outside the interval $0.5 \leq x \leq 4.5$

b.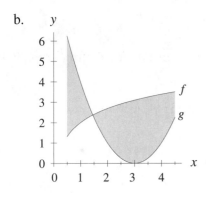

c. $\int_{0.5}^{4.5}[f(x)-g(x)]dx \approx \boxed{4.782}$

d. $\int_{0.5}^{1.458}[g(x)-f(x)]dx + \int_{1.458}^{4.5}[f(x)-g(x)]dx \approx 2.132+6.913 = \boxed{9.045}$

31. a. output interval for b: $-5 < t < 0$
output interval for a: $0 < t < 5$

b. $\int_{-5}^{0}b(t)dt + \int_{0}^{5}a(t)dt = 3590 + 426$; During the 5 months prior to Hurricane Katrina, births in the FEMA-designated assistance counties or parishes within a 100-mile radius of the Hurricane Katrina storm path increased by 3,590; during the 5 months after Hurricane Katrina, births in the FEMA-designated assistance counties or parishes within a 100-mile radius of the Hurricane Katrina storm path increased by only 426.

33. a. $\dfrac{\int_{0}^{12}f(x)dx}{12-0} = \dfrac{1546.14}{12} = \boxed{\$128.845 \text{ billion}}$

b. $\dfrac{f(12)-f(6)}{12-6} = \boxed{2.79 \text{ billion dollars per month}}$

Chapter 6   Analyzing Accumulated Change: Integrals in Action

Section 6.1   Perpetual Accumulation and Improper Integrals (pages 422–423)

1.

$N \to \infty$	$\int_0^N 3e^{-0.2t}\,dt$
5	9.482
25	14.899
125	14.999...
625	15.000...
3125	15.000...
$\lim_{N \to \infty} \int_0^N 3e^{-0.2t}\,dt \approx 15.0$	

$\int_0^\infty 3e^{-0.2t}\,dt \approx \boxed{15.0}$

3.

$N \to -\infty$	$\int_N^3 2e^x\,dx$
−10	40.17098
−40	40.17107
−160	40.17107
−640	40.17107
$\lim_{N \to -\infty} \int_N^3 2e^x\,dx \approx 40.171$	

$\int_{-\infty}^3 2e^x\,dx \approx \boxed{40.171}$

5. $\int_{0.36}^\infty 9.6x^{-0.432}\,dx = \lim_{N \to \infty} \int_{0.36}^N 9.6x^{-0.432}\,dx$

$= \lim_{N \to \infty} \left(\frac{9.6}{0.568} x^{0.568} \Big|_{0.36}^N \right)$

$= \lim_{N \to \infty} \left(\frac{9.6}{0.568} N^{0.568} - \frac{9.6}{0.568}(0.36^{0.568}) \right)$

$= \frac{9.6}{0.568} \lim_{N \to \infty} N^{0.568} - \lim_{N \to \infty} \frac{9.6}{0.568}(0.36^{0.568})$

As $N \to \infty$, $N^{0.568}$ increases without bound: $\lim_{N \to \infty} N^{0.568} = \infty$

$\int_{0.36}^\infty 9.6x^{-0.432}\,dx$ $\boxed{\text{diverges}}$.

7. $\int_5^\infty 5(0.36^x)dx = \lim_{N\to\infty} \int_5^N 5(0.36^x)dx$

$= \lim_{N\to\infty} \left(\frac{5(0.36^x)}{\ln 0.36}\right)\bigg|_5^N$

$= \lim_{N\to\infty} \left(\frac{5(0.36^N)}{\ln 0.36} - \frac{5(0.36^5)}{\ln 0.36}\right)$

$= \frac{5}{\ln 0.36} \lim_{N\to\infty} (0.36^N) - \lim_{N\to\infty} \frac{5(0.36^5)}{\ln 0.36}$

$= \frac{5}{\ln 0.36}(0) - \frac{5(0.36^5)}{\ln 0.36}$

$\approx \boxed{0.0296}$

9. $\int_{10}^\infty 3x^{-2}dx = \lim_{N\to\infty} \int_{10}^N 3x^{-2}dx$

$= \lim_{N\to\infty}\left(-3x^{-1}\bigg|_{10}^N\right)$

$= \lim_{N\to\infty}\left(\frac{-3}{N} - \frac{-3}{10}\right)$

$= -3 \lim_{N\to\infty}\frac{1}{N} + \lim_{N\to\infty}\frac{3}{10}$

$= -3(0) + \frac{3}{10}$

$= \boxed{0.3}$

11. $\int_2^\infty \frac{1}{\sqrt{x}}dx = \lim_{N\to\infty}\int_2^N x^{-0.5}dx$

$= \lim_{N\to\infty}\left(2x^{0.5}\bigg|_2^N\right)$

$= \lim_{N\to\infty}(2\sqrt{N} - 2\sqrt{2})$

$= 2\lim_{N\to\infty}\sqrt{N} - \lim_{N\to\infty} 2\sqrt{2}$

As $N \to \infty$, $\sqrt{N}$ increases without bound: $\lim_{N\to\infty}\sqrt{N} = \infty$

$\int_2^\infty \frac{1}{\sqrt{x}}dx$ $\boxed{\text{diverges}}$.

13. $\displaystyle\int_{-\infty}^{-2}\frac{3}{x^3}\,dx = \lim_{N\to-\infty}\int_{N}^{-2}3x^{-3}\,dx$

$\displaystyle = \lim_{N\to-\infty}\left(-\frac{3}{2}x^{-2}\Big|_{N}^{-2}\right)$

$\displaystyle = \lim_{N\to-\infty}\left(\frac{-3}{2}(-2)^{-2} - \frac{-3}{2}N^{-2}\right)$

$\displaystyle = \lim_{N\to-\infty}\frac{-3}{2}(-2)^{-2} + \frac{3}{2}\lim_{N\to-\infty}N^{-2}$

$\displaystyle = \frac{-3}{2}(-2)^{-2} + \frac{3}{2}(0)$

$\displaystyle = \boxed{-\frac{3}{8}}$

15. $\displaystyle\int_{1}^{\infty}\frac{10}{x}\,dx = \lim_{N\to\infty}\int_{1}^{N}\frac{10}{x}\,dx$

$\displaystyle = \lim_{N\to\infty}(10\ln x)\Big|_{1}^{N}$

$\displaystyle = \lim_{N\to\infty}(10\ln N - 10\ln 1)$

$\displaystyle = 10\lim_{N\to\infty}\ln N - \lim_{N\to\infty}0$

As $N\to\infty$, $\ln N$ increases without bound: $\displaystyle\lim_{N\to\infty}\ln N = \infty$

$\displaystyle\int_{1}^{\infty}\frac{10}{x}\,dx$ $\boxed{\text{diverges}}$.

17. Let $u = x^2 + 1$ and $du = 2x\,dx$
Convert $x = 2$ to $u = 5$

$\displaystyle\int_{2}^{\infty}\frac{2x}{x^2+1}\,dx = \int_{5}^{\infty}\frac{1}{u}\cdot du$

$\displaystyle = \lim_{N\to\infty}\int_{5}^{N}\frac{1}{u}\cdot du$

$\displaystyle = \lim_{N\to\infty}(\ln u)\Big|_{5}^{N}$

$\displaystyle = \lim_{N\to\infty}\ln N - \lim_{N\to\infty}\ln 5$

As $N\to\infty$, $\ln N$ increases without bound: $\displaystyle\lim_{N\to\infty}\ln N = \infty$

$\displaystyle\int_{2}^{\infty}\frac{2x}{x^2+1}\,dx$ $\boxed{\text{diverges}}$.

19. Let $u = x^4 + 1$ and $du = 4x^3 dx$
    Convert $x = 2$ to $u = 17$

$$\int_2^\infty \frac{x^3}{x^4+1} dx = \frac{1}{4} \int_2^\infty \frac{1}{x^4+1} \cdot 4x^3 dx$$

$$= \frac{1}{4} \int_{17}^\infty \frac{1}{u} du$$

$$= \lim_{N \to \infty} \frac{1}{4} \int_{17}^N \frac{1}{u} du$$

$$= \lim_{N \to \infty} \frac{1}{4} \ln u \Big|_{17}^N$$

$$= \lim_{N \to \infty} \frac{1}{4} \ln N - \lim_{N \to \infty} \frac{1}{4} \ln 17$$

As $N \to \infty$, $\ln N$ increases without bound: $\lim_{N \to \infty} \frac{1}{4} \ln N = \infty$

$\int_2^\infty \frac{x^3}{x^4+1} dx$ $\boxed{\text{diverges}}$.

21. $\int_{-\infty}^{-2} \frac{3x^4}{x^6} dx = \int_{-\infty}^{-2} 3x^{-2} dx$

$$= \lim_{N \to -\infty} \int_N^{-2} 3x^{-2} dx$$

$$= \lim_{N \to -\infty} \frac{-3}{x} \Big|_N^{-2}$$

$$= \lim_{N \to -\infty} \left(\frac{-3}{-2} + \frac{3}{N} \right)$$

$$= \lim_{N \to -\infty} \left(\frac{3}{2} \right) + \lim_{N \to -\infty} \left(\frac{3}{N} \right)$$

$$= \frac{3}{2} + 0$$

$$= \boxed{\frac{3}{2}}$$

23. a. The first 1000 years: $\int_0^{1000} -0.027205(0.998188^t) dt \approx \boxed{-12.554 \text{ grams}}$

The fourth 1000 years: $\int_{3000}^{4000} -0.027205(0.998188^t) dt \approx \boxed{-0.054 \text{ grams}}$

b. $\int_0^\infty r(t)dt = \lim_{N\to\infty} \int_0^N r(t)dt$

$= \lim_{N\to\infty} \dfrac{-0.027205(0.998188^t)}{\ln 0.998188} \bigg|_0^N$

$= \lim_{N\to\infty} \dfrac{-0.027205(0.998188^N)}{\ln 0.998188} - \lim_{N\to\infty} \dfrac{-0.027205(0.998188^0)}{\ln 0.998188}$

$= 0 + \dfrac{0.027205}{\ln 0.998188}$

$\approx \boxed{-15.000 \text{ grams}}$

25. a. Change between 25 and 100 years: $\int_{25}^{100} \dfrac{2500}{x^{1.5}} dx = \int_{25}^{100} 2500x^{-1.5} dx = \boxed{\$500}$

After 100 years: $500 + 300 = \boxed{\$800}$

b. $\int_{25}^\infty \dfrac{2500}{x^{1.5}} dx = \lim_{N\to\infty} \int_{25}^N 2500x^{-1.5} dx$

$= \lim_{N\to\infty} \left(-5000 x^{-0.5}\right)\bigg|_{25}^N$

$= \lim_{N\to\infty}\left(-5000 N^{-0.5}\right) - \lim_{N\to\infty}\left(-5000\left(25^{-0.5}\right)\right)$

$= -5000 \lim_{N\to\infty} N^{-0.5} + 1000$

$= 5000(0) + 1000$

$= \boxed{\$1000}$

Section 6.2   Streams in Business and Biology (pages 435–438)

1. a. $R(t) = 1.8(1.03^t)$ million dollars per year

   b. $\int_0^5 1.8(1.03^t) e^{0.0475(5-t)} dt \approx \boxed{\$10.916 \text{ million}}$

   c. $\int_0^5 1.8(1.03^t) e^{-0.0475t} dt \approx \boxed{\$8.608 \text{ million}}$

3. a. $R(t) = (1.8)(0.93^t)$ million dollars per year

   b. $\int_0^5 (1.8)(0.93^t) e^{0.0475(5-t)} dt \approx \boxed{\$8.581 \text{ million}}$

   c. $\int_0^5 (1.8)(0.93^t) e^{-0.0475t} dt \approx \boxed{\$6.767 \text{ million}}$

5. a. $R(t) = (1.8 - 0.04t)$ million dollars per year

b. $\int_0^5 (1.8 - 0.04t)e^{0.0475(5-t)} dt \approx \boxed{\$9.617 \text{ million}}$

c. $\int_0^5 (1.8 - 0.04t)e^{-0.0475t} dt \approx \boxed{\$7.584 \text{ million}}$

7. a. i. Aligning input to $t = 0$ in the current year gives: input values $\{-5, -4, -3, -2, -1\}$

$\boxed{P_a(t) \approx 5.714t^2 + 82.286t + 1128}$ thousand dollars per year

ii. $\int_0^3 0.5 \cdot P_a(t)e^{0.064(3-t)} dt \approx \boxed{\$2089.927 \text{ thousand}}$

b. i. Last year's percentage change: $\dfrac{1130 - 1050}{1050} \cdot 100\% = 7.6\%$

$\boxed{P_b(t) = 1130(1.076^t)}$ thousand dollars per year

ii. $\int_0^3 0.5 \cdot P_b(t)e^{0.064(3-t)} dt \approx \boxed{\$2082.544 \text{ thousand}}$

c. i. $P_c(t) = 1130$ thousand dollars per year

ii. $\int_0^3 0.5 \cdot P_c(t)e^{0.064(3-t)} dt \approx \boxed{\$1868.654 \text{ thousand}}$

d. i. increase over last year: $1130 - 1050 = 80$ thousand dollars

$\boxed{P_d(t) = 1130 + 80t}$ thousand dollars per year

ii. $\int_0^3 0.5 \cdot P_d(t)e^{0.064(3-t)} dt \approx \boxed{\$2060.749 \text{ thousand}}$

9. a. $R(t) = 0.10(182.52)(1.05^t)$ billion dollars per year

b. $\int_0^9 0.10(182.52)(1.05^t)e^{-0.048t} dt \approx \boxed{\$164.853 \text{ billion}}$

11. a. $\int_0^5 48.23 e^{0.056(5-t)} dt \approx \boxed{\$278.296 \text{ billion}}$

b. $\int_0^5 48.23 e^{-0.056t} dt \approx \boxed{\$210.331 \text{ billion}}$

13. a. $\int_0^8 R(t)e^{0.062(8-t)} dt \approx \boxed{\$806.877 \text{ million}}$

b. $\int_0^8 R(t)e^{-0.062t} dt \approx \boxed{\$491.357 \text{ million}}$

15. a. flow rate: $R(t) = 0.273(1.1^t)$ billion dollars per year

present value: $\int_0^{15} 0.273(1.1^t)e^{-0.13t} dt \approx \boxed{\$3.193 \text{ billion}}$

b. flow rate: $R(t) = \$0$ per year

present value: $\int_0^{15} 0.273 e^{-0.15t} dt \approx \boxed{\$1.628 \text{ billion}}$

c. Answers will vary but should be similar to the following:

According to the assumptions given, Tenet purchased the company for less than they believed the company was worth, and the purchase price was more than Ornda believed the company was worth. (Note: These answers assume that the annual returns given refer to APRs, compounded continuously. If they are interpreted as APYs, the answers to parts *a* and *b* are $3.369 billion and $1.713 billion, respectively.)

17. a. flow rate: $R(t) = 1.4(1.05^t)$ million dollars per year

   present value: $\int_0^{10} 1.4(1.05^t)e^{-0.08t}\,dt \approx$ $12.03 million

   b. flow rate: $R(t) = 2.8$ million dollars per year

   present value: $\int_0^{10} 2.8 e^{-0.08t}\,dt \approx$ $19.27 million

   c. Answers will vary but should include the following information:
   Company A is willing to offer up to $12.03 million, but Company B demands at least $19.27 million.

19. flow rate: $R(t) = 1.2(1.03^t)$ million dollars per year

   present value: $\int_0^5 1.2(1.03^t)e^{-0.06t}\,dt \approx$ $5.566 million

21. a. $19{,}000(1 - 0.178)^{30} \approx$ 53 elephants

   b. renewal rate: $r(t) = 47(1 - 0.13)^t = 47(0.87^t)$ elephants per year
   survival rate: $s = 1 - 0.178 = 0.822$
   growth rate: $f(t) = 47(0.87^t)(0.822)^{30-t}$ elephants per year

   c. population:
   $$Ps^b + \int_0^b r(t)s^{b-t}\,dt = 53 + \int_0^{30} 47(0.87^t)(0.822)^{30-t}\,dt$$
   $$\approx 63 \text{ elephants}$$

23. a. $15{,}000(0.75^{21}) \approx$ 36 muskrats

   b. renewal rate: $r(t) = 468$ muskrats per year
   survival rate: $s = 0.75$
   growth rate: $m(t) = 468(0.75)^{21-t}$ muskrats born t years after 1936.

   c. $Ps^b + \int_0^b r(t)s^{b-t}\,dt = 15{,}000(0.75^{21}) + \int_0^{21} 468(0.75)^{21-t}\,dt$
   $\approx$ 1659 muskrats

Section 6.3 Calculus in Economics—Demand and Elasticity (pages 447–451)

1. a. Input units: dollars per flower
   Output units: million flowers
   b. (The answers in the answer key are incorrect.)

   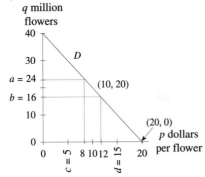

   When demand is 24 million flowers, the price is $8 per flower.
   When demand is 16 million flowers, the price is $12 per flower.

   c.

   At a price of $5 per flower, demand is 30 million flowers.
   At a price of $15 per flower, demand is 10 million flowers.

3. a. Input: dollars per card
   Output: billion cards
   b.

   When demand is 6 billion cards, price is $1.20 per card.
   When demand is 10 billion cards, price is $0.50 per card.

c.

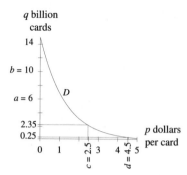

At a price of $2.5 per card, demand is 2.25 billion cards.
At a price of $4.5 per card, demand is 0.25 billion cards.

5. a. Input: cents per pound
   Output: million pounds
   b. At $0.16 per pound, demand is 5 million pounds of rice.
   At $1.60 per pound, demand is 50 thousand pounds of rice.

7. a. Input: dollars per pound
   Output: million pounds
   b. At a price of $5 per pound, demand is 16 million pounds of coffee.
   At a price of $15 per pound, demand is 3 million pounds of coffee.

9. a. Input: dollars per bulb
   Output: million light bulbs
   b. At a price of $1.56 per bulb, demand is 2.49 million light bulbs.
   At a price of $3.65 per bulb, demand is 1.56 million light bulbs.

11. a. Consumer expenditure is shaded on the graph.

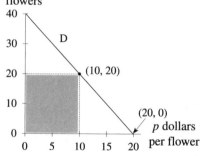

$$\text{area} = (20 \text{ million flowers}) \cdot \left(10 \frac{\text{dollars}}{\text{flower}}\right)$$

$$= \boxed{200 \text{ million dollars}}$$

b. Consumer surplus is shaded on the graph.

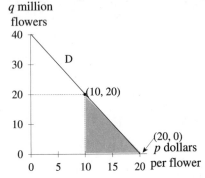

$$\text{area} = \frac{1}{2}(20 \text{ million flowers}) \cdot \left(10 \frac{\text{dollars}}{\text{flower}}\right)$$
$$= \boxed{100 \text{ million dollars}}$$

c. Consumer willingness and ability to spend is the sum of consumer expenditure and consumer surplus: $200 + 100 = \boxed{\$300 \text{ million}}$

13. a. area of the rectangle $= (4.7 \text{ billion cards}) \cdot \left(1.5 \frac{\text{dollars}}{\text{card}}\right)$
$$= \boxed{\$7.05 \text{ billion}}$$

b. The area of the region to the right of $p = 1.5$ is given as $\boxed{\$6.03 \text{ billion}}$.

c. Consumer willingness and ability to spend is the sum of consumer expenditure and consumer surplus: $\boxed{\$13.08 \text{ billion}}$.

15. $D(p) = 50 - 2p$ units
Solving $D(p) = 0$ yields $p = \boxed{25 \text{ dollars per unit}}$

17. $D(p) = 35 - 7\ln p$ units
Solving $D(p) = 0$ yields $p = \boxed{\$148.41 \text{ per unit}}$

19. $D(p) = 3.6 p^{-0.8}$ units
$D(p) = 0$ has no solution; price per unit is $\boxed{\text{unbounded}}$.

21. a. $q_0 = \boxed{5 \text{ thousand units}}$
b. Market price appears to be $\boxed{\text{unbounded}}$.

c.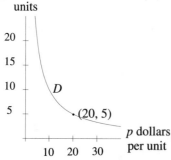

23. a. Solving $D(p)=0$ yields $p = \boxed{\$555 \text{ per chair}}$

According to the model, consumers will no longer purchase chairs at prices of $555 or higher. Because a demand schedule is a model of aggregate behavior, it cannot guarantee individual behavior.

$D(99.95) \approx \boxed{4.550 \text{ million chairs}}$

a. Solving $D(p) = 3$ yields $p_0 = \boxed{\$255 \text{ per chair}}$

b. $\int_{255}^{555} D(p)dp = \boxed{\$450 \text{ million}}$

25. a. $D(p) = 1520.417(0.15003^p)$ million bottles at a price of $p per bottle

$D(p) = 0$ has no solution. There is not a price above which consumers will purchase no bottles of water.

b. $D(2.59) \approx \boxed{11.17 \text{ million bottles}}$

c. Find the amount that consumers are willing and able to spend to purchase the quantity found in part b.

$$p_0 q_0 + \int_{p_0}^{\infty} D(p)dp \approx (2.59)(11.17) + \lim_{N \to \infty} \int_{2.59}^{N} 1520.417\left(0.15003^p\right)dp$$

$$\approx \boxed{\$34.8 \text{ million}}$$

d. Consumer surplus = $\lim_{N \to \infty} \int_{2.59}^{N} 1520.417\left(0.15003^p\right)dp \approx \boxed{\$5.89 \text{ million}}$

27. a. $\eta = \dfrac{p \cdot D'(p)}{D(p)}$

$= \dfrac{-0.01p}{-0.01p + 5.55}$

Solving $\eta = -1$ yields $p = \$277.50$

$D(277.50) = 2.775$ million chairs

The point of unit elasticity is $(\$277.50,\ 2.775 \text{ million chairs})$.

b. Testing elasticity for one point on each side of $p = 277.50$ yields

$$p = 270: \eta = \frac{270 \cdot D'(270)}{D(270)} \approx -0.947 \quad \rightarrow \quad \text{demand is inelastic for} \quad 0 < p < 277.5$$

$$p = 280: \eta = \frac{280 \cdot D'(280)}{D(280)} \approx -1.018 \quad \rightarrow \quad \text{demand is elastic for} \quad 277.5 < p < 555$$

29. a. $\eta = \dfrac{p \cdot D'(p)}{D(p)}$

$= \dfrac{p \cdot (0.06p - 1.6)}{0.03p^2 - 1.6p + 21}$

b. $\eta = \dfrac{15 D'(15)}{D(15)} = -2.8 \quad \rightarrow \quad$ demand is elastic at $p = \$15$

c. At $15 per ticket, a small increase in market price results in a relatively large decrease in demand.

Section 6.4 Calculus in Economics—Supply and Equilibrium (pages 461–464)

1. a. input units: dollars per pot
   output units: million pots

   b.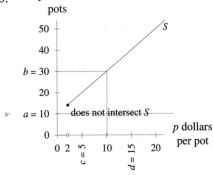

   There is no corresponding market price at which 10 million pots are supplied.
   When supply is 30 million pots, price is $10 per pot.

   c.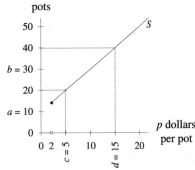

   At a price of $5 per pot, supply is 20 million pots.

At a price of $15 per pot, supply is 40 million pots.

3. a. input units: dollars per card
   output units: billion cards

   b.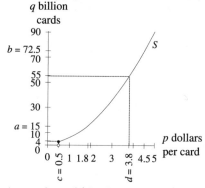

   When supply is 15 billion cards, the price is $1.80 per card.
   When supply is 72.5 billion cards, the price is $4.50 per card.

   c.

   At a price of $0.50 per card, producers are willing to supply about 4 billion cards.
   At a price of $3.80 per card, producers are willing to supply about 55 billion cards.

5. a. input: dollars per pizza
   output: hundred pizzas

   b. At a price of $5 per pizza, supply is 1,600 pizzas.
   At a price of $16 per pizza, supply is 2,400 pizzas.

7. a. input: dollars per pound
   output: million pounds

   b. At a price of $5 per pound, coffee bean supply is 9 million pounds.
   At a price of $15 per pound, coffee bean supply is 40 million pounds.

9. a. Producer revenue is shaded on the graph.

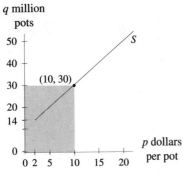

area $= q_0 \cdot p_0 = (30 \text{ million pots}) \cdot \left(10 \dfrac{\text{dollars}}{\text{pot}}\right) = \boxed{\$300 \text{ million}}$

b. Producer surplus is shaded on the graph.

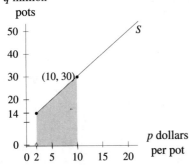

$$\int_{p_s}^{p_0} S(p)dp = \int_{2}^{30} S(p)dp$$
$$= \text{area of the trapezoid}$$
$$= \frac{1}{2}(14+30) \cdot 8$$
$$= \boxed{\$176 \text{ million}}$$

c. Producer willingness and ability to receive is the difference in producer revenue and producer surplus: $300 - 176 = \boxed{\$124 \text{ million}}$

11.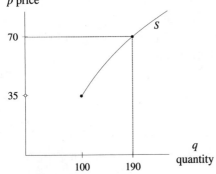

a. 190 units
b. ($35 per unit, 100 units)

c.

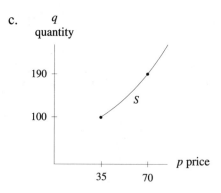

13. a. $4,000 per saddle: $S(4) = $ 0 saddles
       $8,000 per saddle: $S(8) \approx 17.353$ → 17,353 saddles
    b. Solving $S(p) = 10$ yields $p = 5.867$ → $5,867 per saddle
    c. producer revenue : $7.5 \cdot S(7.5) \approx$ 114.373 million dollars
    d. producer surplus : $\int_{5}^{7.5} S(p)dp \approx$ $28.080 million

15. a. $S(p) = \begin{cases} 0 & \text{when } p < 5.00 \\ 0.2p & \text{when } p \geq 5.00 \end{cases}$ gives the quantity of DVDs producers are willing to supply at a price of p dollars per DVD. $S(p)$ is measured in million DVDs.
    b. $S(15.98) \approx$ 3.196 million DVDs
    c. Solving $0.2p = 2.3$ yields $p =$ $11.50 per DVD
    d. producer revenue: $19.99 \cdot S(19.99) \approx$ $79.920 million
       producer surplus: $\int_{5}^{19.99} S(p)dp \approx$ $37.460 million

17. a. $S(9) \approx 6.047$
       shutdown point: ($9.00 per unit, 6.047 million units)
       Producers are willing to supply 6.047 million units at a market price of $9.00 per unit.
       Below $9.00 per unit, producers will shutdown.
    b. Solving $D(p) = S(p)$ yields $p \approx 19.853$
       $D(19.853) = S(19.853) \approx 14.082$
       equilibrium point: ($19.85 per unit, 14.082 million units)
       At the price of $19.85 per unit, producers are willing to buy exactly what consumers are willing to purchase—14.082 million units.

19. a. $S(0.2) = 1.4$
       shutdown point: ($0.20 per unit , 1.4 thousand units)
       Producers are not willing to supply any units at a price of less than $0.20 per unit. At $0.20 per unit, producers are willing to supply 1.4 thousand units.

b. Solving $D(p) = S(p) \rightarrow p = 0.69$ yields
$D(0.69) = S(0.69) = 4.838$
equilibrium point: ($0.69 per unit, 4.838 thousand units)
At the price of $0.69 per unit, producers are willing to buy exactly what consumers are willing to purchase—4.838 thousand units.

21. a. q thousand filters

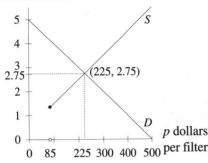

From the graph, $S(p) = D(p)$ when $p =$ $225 per filter
Market equilibrium occurs when 2.75 thousand filters are supplied and demanded at a price of $225 per filter.

b. q thousand filters

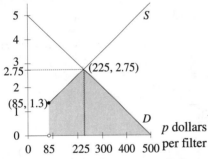

Producer surplus is represented by the trapezoid from $p = 85$ to $p = 225$:
$$\text{area} = \frac{1}{2}(1.3 + 2.75) \cdot (225 - 85) = \$283.5 \text{ thousand}$$
Consumer surplus is represented by the triangle from $p = 225$ to $p = 500$:
$$\text{area} = \frac{1}{2} 2.75 \cdot (500 - 225) = \$378.125 \text{ thousand}$$
Total social gain is calculated as $283.5 + 378.125 =$ $661.625 thousand

23. a. $S(1.5) \approx$ 1.653 million pounds
$D(1.5) \approx 37.881$ million pounds $> S(1.5)$
At a price of $1.50 per pound the demand exceeds the supply.

b. Solving $S(p) = D(p)$ yields $p \approx 9.26$
$S(9.26) = D(9.26) \approx 21.32$
market equilibrium: ($9.26 per pound, 21.32 million pounds)

c. consumer surplus: $\int_{9.26}^{\infty} D(p)dp = \lim_{N \to \infty} \int_{9.26}^{N} D(p)dp \approx \$86.113$ million

producer surplus: $\int_{0.5}^{9.26} S(p)dp \approx \$67.010$ million

total social gain: $86.113 + 67.010 \approx \$153.123$ million.

25. a. $D(p) = \dfrac{38.301}{1 + 0.003e^{0.050p}}$ gives demand and $S(p) = \begin{cases} 0 & \text{when } p < 47.5 \\ 0.747p - 35.467 & \text{when } p \geq 47.5 \end{cases}$ gives supply when \$p is the price of one calculator, $60 \leq p \leq 210$. The output values of both demand and supply are measured in million calculators.

b. Solving $D(p) = S(p)$ yields $p \approx 87.788$ → $\boxed{\$87.79 \text{ per calculator}}$

$D(87.788) = S(87.788) \approx \boxed{30.08 \text{ million calculators}}$

c. producer surplus: $\int_{47.50}^{87.79} S(p)dp \approx \boxed{\$606.026 \text{ million}}$

consumer surplus: $\int_{87.79}^{\infty} D(p)dp = \lim_{N \to \infty} \int_{87.79}^{N} D(p)dp \approx \boxed{\$1184.337 \text{ million}}$

total social gain: $606.026 + 1184.337 \approx \boxed{\$1790.363 \text{ million}}$

Section 6.5   Calculus in Probability—Part 1 (pages 472–475)

1. There is a 46% chance that any telephone call made on a computer software technical support line will be 5 minutes or more.

3. In March, there is a 15% likelihood that New Orleans will receive between 2 and 4 inches of rain.

5. There is a 25% chance that any car rented from Hertz at the Los Angeles airport on 12/28/2013 is at least two years old.

7. $h(x)$ is normally distributed with a mean of 15 and a smaller standard deviation than in Activity 8 so the narrower distribution is correct.

9. $g(x)$ shows that there is an equal probability that any number from 0 to 100 will be chosen.

11. $t(x)$ shows that the probability of the random variable occurring decreases as the random variable increases.

13. a. $0.004 + 0.004 = 0.008$
    b. $0.124 + 0.194 = 0.318$; Approximately 31.8% of females who contracted cancer were between 40 and 60 years old.

15. no; $\int_{-1}^{3} f(x)dx = 3$ which is greater than 1.

17. no; $h(x)$ is not always positive on the interval $0 < x < 1$.

19. yes; $R(x) \geq 0$ for each real number x and

$$\int_{-\infty}^{\infty} R(x)dx = \int_{1}^{2} R(x)dx + \int_{2}^{3} R(x)dx$$
$$= \frac{1}{2}(1)(0.5+1) + (1)(0.25)$$
$$= 1$$

21. $f(w) \geq 0$ for each real number w and

$$\int_{0}^{1} f(w)dw = 1.5 \int_{0}^{1} (1-w^2) dw$$
$$= 1.5 \left(x - \frac{1}{3}x^3 \right) \Big|_{0}^{1}$$
$$= 1.5 \left(1 - \frac{1}{3} \right)$$
$$= 1$$

Yes, $f(w)$ is a valid probability density function.

23. no; $\int_{-\infty}^{\infty} p(t)dt = \int_{1}^{3} \frac{\ln 3}{t} dt \approx 1.207$

25. no; $\int_{-\infty}^{\infty} 0.625 e^{-1.6y} dy$

$$= \lim_{N \to \infty} \int_{0}^{N} 0.625 e^{-1.6y} dy$$
$$= \lim_{N \to \infty} \frac{0.625 e^{-1.6y}}{-1.6} \Big|_{0}^{N}$$
$$= \lim_{N \to \infty} \frac{0.625 e^{-1.6N}}{-1.6} - \lim_{N \to \infty} \frac{0.625 e^{-1.6(0)}}{-1.6}$$
$$= \frac{0.625}{-1.6} \lim_{N \to \infty} e^{-1.6N} + \lim_{N \to \infty} \frac{0.625}{1.6}$$
$$= \frac{0.625}{-1.6}(0) + \frac{0.625}{1.6}$$
$$= .391$$

27. a. $\int_{0}^{1} 0.32x\,dx = 0.16$

b. $\mu = \int_0^{2.5} x(0.32x)dx$

$= \int_0^{2.5} 0.32x^2 dx$

$\approx 1.667 \quad \rightarrow \quad 167$ gallons

c.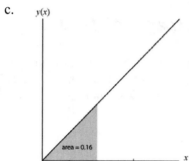

29. a. $\mu = \int_0^4 t\left(\frac{3}{32}(4t - t^2)\right) dt$

$= \int_0^4 \left(\frac{3}{8}t^2 - \frac{3}{32}t^3\right) dt$

$= \left(\frac{3t^3}{24} - \frac{3t^4}{128}\right)\Big|_0^4$

$= \left(\frac{3(4)^3}{24} - \frac{3(4)^4}{128}\right) - \left(\frac{3(0)^3}{24} - \frac{3(0)^4}{128}\right)$

$= 2$ minutes

b. $\sigma = \sqrt{\int_0^4 (t-2)^2 \cdot \frac{3}{32}(4t - t^2) dt}$

≈ 0.894 minutes

c. $\int_0^3 g(t)dt = 0.844$

There is an 84% chance that the time it takes a child between the ages of 8 and 10 to learn the rules of this game is less than 3 minutes.

$\mu = \int_0^\infty x(0.1163e^{-0.1163x}) dx$

$= \lim_{N \to \infty} \int_0^N x(0.1163e^{-0.1163x}) dx$

$= 8.598$ hours

31. The bar graph is not a probability density function because it is not continuous. Furthermore, the sum of the shaded areas is greater than 1.

33. Answers will vary but might be similar to the following:
The probability is defined as a portion of an integral of a positive function. The integral will never be negative since the function has no negative values. Because of the condition of the definition that $\int_{-\infty}^{\infty} f(x)dx = 1$, the probability will range between 0 and 1.

Section 6.6   Calculus in Probability—Part 2 (pages 482–486)

1. true; The left end behavior of any cumulative distribution function will be 0, corresponding to impossible events. The right end behavior of any cumulative distribution function will be 1, corresponding to sure events. Cumulative distribution functions are always non-decreasing.

3. false; The value that makes $G(t) = \begin{cases} ke^{-t} & \text{when } t \geq 0 \\ 0 & \text{when } t < 0 \end{cases}$ is $k = 1$.

5. yes; $G(t)$ could be a cumulative distribution function because it is non-decreasing.

7. no; $S(x)$ could not be a cumulative distribution function because the left end behavior appears to only approach, but not reach a value of 0.

9. a. The uniform density function for the wait time is $s(t) = \begin{cases} \dfrac{1}{30} & \text{when } 0 \leq t \leq 30 \\ 0 & \text{elsewhere} \end{cases}$

   So, $P(t \geq 10) = P(10 \leq t \leq 30) = \dfrac{1}{30} \cdot 20 = 0.667$

   66.7%

   b. $P(t \leq 20) = P(0 \leq t \leq 20) = \dfrac{1}{30} \cdot 20 = 0.667$

   66.7%

   c. $\mu = \int_{-\infty}^{\infty} t \cdot s(t) dt$

   $= \int_{0}^{30} \left(t \cdot \dfrac{1}{30} \right) dt$

   $= \dfrac{t^2}{60} \Big|_{0}^{30}$

   $= \dfrac{30^2}{60} - 0$

   $= 15$ seconds

11. a. The exponential distribution is $E(t) = 0.4e^{-0.4t}$ when $t \geq 0$, where t is the waiting time in minutes.

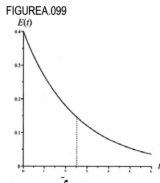
FIGURE A.099

b. $P(t < 2) = \int_{-\infty}^{2} E(t)\,dt$

$= \int_{0}^{2} 0.4e^{-0.4t}\,dt$

$= \left. \dfrac{0.4e^{-0.4t}}{-0.4} \right|_{0}^{2}$

$= -e^{-0.4(2)} - (-e^{-0.4(0)})$

≈ 0.551

55.1%

c. $P(2 < t < 4) = \int_{2}^{4} E(t)\,dt$

$= \int_{2}^{4} 0.4e^{-0.4t}\,dt$

$= \left. \dfrac{0.4e^{-0.4t}}{-0.4} \right|_{2}^{4}$

$= -e^{-0.4(4)} - (-e^{-0.4(2)})$

≈ 0.247

24.7%

d. $P(t > 5) = \int_5^\infty E(t)dt$

$= \lim_{N \to \infty} \int_5^N 0.4e^{-0.4t} dt$

$= \lim_{N \to \infty} \dfrac{0.4e^{-0.4t}}{-0.4} \Big|_5^N$

$= \lim_{N \to \infty} -e^{-0.4t} \Big|_5^N$

$= \lim_{N \to \infty} \left[-e^{-0.4N} - \left(-e^{-0.4(5)} \right) \right]$

$= \lim_{N \to \infty} \left[-e^{-0.4N} - \left(-e^{-0.4(5)} \right) \right]$

$= 0 + e^{-2}$

$= 0.135$

13.5%

13. a. $f(x) = \dfrac{1}{10.63\sqrt{2\pi}} e^{\dfrac{-(x-40)^2}{2(10.63^2)}}$

$P(0 \le x \le 45) = \int_0^{45} f(x)dx \approx 0.681$

The probability that a piece of luggage weighs less than 45 pounds is approximately 68.1%.

b. In this case, $\mu = (40)(80) = 3200$ and $\sigma = (10.63)(80) = 850.4$.

$f(x) = \dfrac{1}{850.4\sqrt{2\pi}} e^{\dfrac{-(x-3200)^2}{2(850.4^2)}}$

$P(1200 \le x \le 2400) = \int_{1200}^{2400} f(x)dx \approx 0.164$

The probability that the total weight of the luggage for 80 passengers is between 1200 and 2400 pounds is approximately 16.4%.

c. The probability density function for the weight of a passenger's luggage is decreasing most rapidly one standard deviation to the right of the mean (50.63 pounds).

15. $f(x) = \dfrac{1}{8.372\sqrt{2\pi}} e^{\dfrac{-(x-5.3)^2}{2(8.372^2)}}$

a. $\mu - \sigma = -3.072$ and $\mu + \sigma = 13.672$:

$\int_{-3.072}^{13.672} f(x)dx \approx 0.68$

$\mu - 2\sigma = -11.444$ and $\mu + 2\sigma = 22.044$:

$\int_{-11.444}^{22.044} f(x)dx \approx 0.95$

$\mu - 3\sigma = -19.816$ and $\mu + 3\sigma = 30.416$:

$$\int_{-19.816}^{30.416} f(x)dx \approx 0.997$$

b. 0.815
c. 0.818

17. The peak of the second normal distribution is higher.

19. $f(x) = \dfrac{1}{28.65\sqrt{2\pi}} e^{\frac{-(x-72.3)^2}{2(28.65^2)}}$

   a. $P(60 \le x \le 80) = \int_{60}^{80} f(x)dx \approx 0.272$
   Approximately 27.2% of the students are likely to make a score between 60 and 80.

   b. $P(x > 90) = 0.5 - \int_{72.3}^{90} f(x)dx \approx 0.268$
   Approximately 26.8% of the students are likely to make a score of at least 90.

   c. $P(x < 43.65$ or $x > 100.95) = 1 - P(43.65 \le x \le 100.95)$
   $= 1 - \int_{43.65}^{100.95} f(x)dx \approx 0.317$
   Approximately 31.7% of the students made a score that was more than one standard deviation away from the mean.

   d. The rate of change is a maximum at one standard deviation less than the mean, $72.3 - 28.65 = 43.65$.

21.

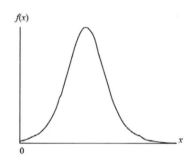

23. a. $G(x) = \begin{cases} 0 & \text{when } x < 0 \\ 1 - e^{-2x} & \text{when } x \ge 0 \end{cases}$

   b. Using g: $P(x \le 0.35) = \int_{-\infty}^{.35} g(x)dx \approx 0.503$
   Using G: $P(x < 0.35) = G(0.35) \approx 0.503$

   c. 0.179

25. a. $F(x) = \begin{cases} 0 & \text{when } x < 0 \\ x^2 & \text{when } 0 \leq x < 1 \\ 1 & \text{when } x \geq 1 \end{cases}$

b. Using f: $P(x < 0.67) = \int_{-\infty}^{0.67} f(x)dx \approx 0.449$

Using F: $P(x < 0.67) = F(0.67) \approx 0.449$

Section 6.7 Differential Equations—Slope Fields and Solutions
(pages 494–498)

1. $\dfrac{dy}{dx} = 2x$

   $dy = 2x\,dx$

   $\int dy = \int 2x\,dx$

   $\boxed{y = x^2 + C}$

3. $\dfrac{dy}{dx} = 6x^2 y$

   $\dfrac{dy}{y} = 6x^2\,dx$

   $\int \dfrac{dy}{y} = \int 6x^2\,dx$

   $\ln|y| = \dfrac{6x^3}{3} + C$

   $e^{\ln|y|} = e^{2x^3 + C}$

   $\boxed{y = ae^{2x^3}}$

5. $\dfrac{dy}{dx} = -x$

   $dy = -x\,dx$

   $\int dy = \int -x\,dx$

   $\boxed{y = \dfrac{-x^2}{2} + C}$

7. $\dfrac{dy}{dx} = 10xy^{-1}$

   $y\,dy = 10x\,dx$

   $\int y\,dy = \int 10x\,dx$

   $\dfrac{y^2}{2} = \dfrac{10x^2}{2} + C_1$

   $\boxed{y^2 = 10x^2 + C_2}$

9.
$$\frac{dy}{dx} = e^{0.05x} e^{-0.05y}$$

$$\frac{dy}{e^{-0.05y}} = e^{0.05x} dx$$

$$e^{0.05y} dy = e^{0.05x} dx$$

$$\int e^{0.05y} dy = \int e^{0.05x} dx$$

$$\frac{e^{0.05y}}{0.05} = \frac{e^{0.05x}}{0.05} + C_1$$

$$\boxed{e^{0.05y} = e^{0.05x} + C_2}$$

11.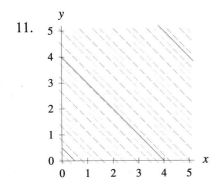

13.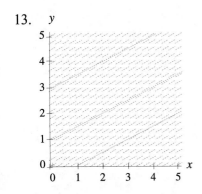

15.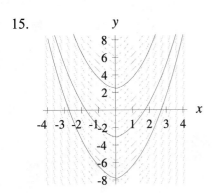

17.
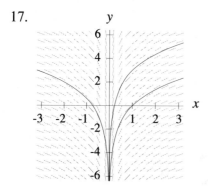

19.
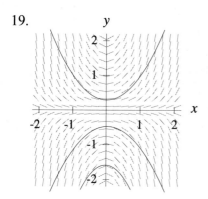

21.
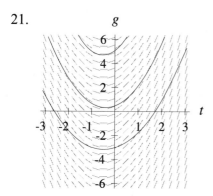

23.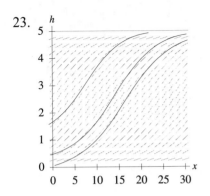

25.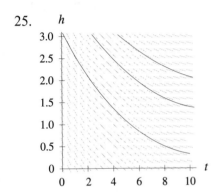

27.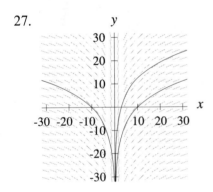

29.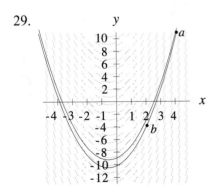

31.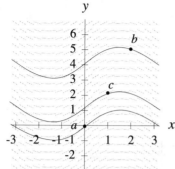

33.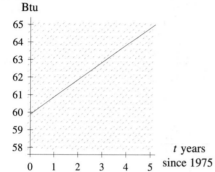

35. Initial condition (1, 4); $\Delta x = \dfrac{7-1}{2} = 3$

x	Estimate of $y(x)$ $y_i = y_{i-1} + \dfrac{dy_{i-1}}{dx}\Delta x_{i-1}$	$\dfrac{dy}{dx} = 2x$
1	4	2
4	$4 + 2 \cdot 3 = 10$	8
7	$10 + 8 \cdot 3 = \boxed{34}$	

37. Initial condition (2, 2); $\Delta x = \dfrac{8-2}{2} = 3$

x	Estimate of $y(x)$ $y_i = y_{i-1} + \dfrac{dy_{i-1}}{dx}\Delta x_{i-1}$	$\dfrac{dy}{dx} = \dfrac{5}{x}$
2	2	2.5
5	$2 + 2.5 \cdot 3 = 9.5$	1
8	$9.5 + 1 \cdot 3 = \boxed{12.5}$	

39. a. Initial condition (1, 6); $\Delta x = 0.25$ month

t	Estimate of $w(t)$	$w'(t)$
1	6	33.68
1.25	$6 + 33.68 \cdot 0.25 = 14.42$	26.944
1.5	$14.42 + 26.944 \cdot 0.25 = 21.156$	22.453
1.75	$21.156 + 22.453 \cdot 0.25 \approx 26.769$	19.246
2	$26.769 + 19.246 \cdot 0.25 \approx 31.581$	16.84
2.25	$31.581 + 16.840 \cdot 0.25 \approx 35.791$	14.969
2.5	$35.791 + 14.969 \cdot 0.25 \approx 39.533$	13.472
2.75	$39.533 + 13.472 \cdot 0.25 \approx 42.901$	12.247
3	$42.901 + 12.247 \cdot 0.25 \approx \boxed{45.963 \text{ pounds}}$	11.227
3.25	$45.963 + 11.227 \cdot 0.25 \approx 48.769$	10.363
3.5	$48.769 + 10.363 \cdot 0.25 \approx 51.360$	9.623
3.75	$51.360 + 9.623 \cdot 0.25 \approx 53.766$	8.981
4	$53.766 + 8.981 \cdot 0.25 \approx 56.011$	8.42
4.25	$56.011 + 8.420 \cdot 0.25 \approx 58.116$	7.925
4.5	$58.116 + 7.925 \cdot 0.25 \approx 60.097$	7.484
4.75	$60.097 + 7.484 \cdot 0.25 \approx 61.969$	7.091
5	$61.969 + 7.091 \cdot 0.25 \approx 63.741$	6.736
5.25	$63.741 + 6.736 \cdot 0.25 \approx 65.425$	6.415
5.5	$65.425 + 6.415 \cdot 0.25 \approx 67.029$	6.124
5.75	$67.029 + 6.124 \cdot 0.25 \approx 68.560$	5.857
6	$68.560 + 5.857 \cdot 0.25 \approx \boxed{70.024 \text{ pounds}}$	

b. Initial condition (1, 6); $\Delta x = 1$ month

t	Estimate of $w(t)$	$w'(t)$
1	6	33.68
2	$6 + 33.68 \cdot 1 = 39.68$	16.84
3	$39.68 + 16.84 \cdot 1 = \boxed{56.52 \text{ pounds}}$	11.227
4	$56.52 + 11.227 \cdot 1 \approx 67.747$	8.42
5	$67.747 + 8.42 \cdot 1 \approx 76.167$	6.736
6	$76.167 + 6.736 \cdot 1 \approx \boxed{82.903 \text{ pounds}}$	

41. a. $\dfrac{dp}{dt} = 3.9t^{3.55} e^{-1.351t}$ thousand barrels per year t years after production begins, where $p(t)$ is the total amount of oil produced after t years

b. Initial condition: (0, 0); $\Delta x = \dfrac{5}{10} = 0.5$ year

t	Estimate of $p(t)$	$p'(t)$
0	0	0
0.5	0	0.087
1	0.043	1.017
1.5	0.552	4.290
2	2.697	11.911
2.5	8.652	26.302
3	21.803	50.244
3.5	46.925	86.845
4	90.348	139.513
4.5	160.104	211.937
5	266.073	

After 5 years of production, the well has produced approximately 266.073 thousand barrels.

c. $\dfrac{dp}{dt}$ thousand barrels per year

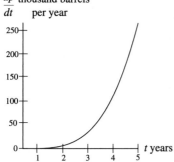

Section 6.8 Differential Equations—Proportionality and Common Forms (pages 507–512)

1. $c = kg$

3. $\dfrac{dp}{da} = ka$

5. $\dfrac{dT}{dt} = \dfrac{k}{T}$

7. $\dfrac{dA}{dt} = kA$

9. $\dfrac{dx}{dt} = kx(N - x)$

11. Flow-in rate: $\dfrac{k}{\sqrt{D}}$ where k is a constant

    Flow-out rate: hD where h is a constant

    $$\boxed{\dfrac{dD}{dt} = \dfrac{k}{\sqrt{D}} - hD}$$

13. a. $\dfrac{dc}{dt} = 1.08$ quadrillion Btu per year

    b. $\dfrac{dc}{dt} = 1.08$

    $\int dc = \int 1.08 \, dt$

    $\boxed{c(t) = 1.08t + K \text{ quadrillion Btu}}$

    c. Solving $c(5) = 76.0$ yields $K = 70.6$

    $\boxed{c(t) = 1.08t + 70.6 \text{ quadrillion Btu}}$

    d. $c(0) = \boxed{70.6 \text{ quadrillion Btu}}$

    $\left.\dfrac{dc}{dt}\right|_0 = \boxed{1.08 \text{ quadrillion Btu per year}}$

6.8 Differential Equations—Proportionality and Common Forms

Solutions to Odd-Numbered Activities

15. a. $\dfrac{dw}{dt} = \dfrac{k}{t}$ pounds per month

  b. $\dfrac{dw}{dt} = \dfrac{k}{t}$

  $\int dw = \int \dfrac{k}{t} dt$

  $w(t) = k \ln t + C$

  Solving $w(1) = 6$ yields $C = 6$

  Solving $w(9) = 80$, so $k \approx 33.679$

  $\boxed{w(t) \approx 33.679 \ln t + 6 \text{ pounds}}$

  c. $w(3) \approx \boxed{43 \text{ pounds}}$

  $w(6) \approx \boxed{66 \text{ pounds}}$

  d. Answers will vary but may include the following information.
  The rate of increase of the weight of the dog should slow and eventually become zero as the dog reaches adulthood. However, the differential equation indicates that the rate increases infinitely.

17. a. $\dfrac{dq}{dt} = kq$ milligrams per hour

  b. $\dfrac{dq}{dt} = kq$

  $\int \dfrac{1}{q} dq = \int k\, dt$

  $\ln|q| + c_1 = kt + c_2$

  $\ln|q| = kt + C$

  $q = e^{kt+C}$

  $q(t) = ae^{kt}$

  Solving $q(0) = 200$ yields $a = 200$

  Solving $q(2) = 100$ yields $k \approx -0.347$

  $\boxed{q(t) \approx 200 e^{-0.347t} \text{ milligrams}}$

  c. $q(4) \approx \boxed{50 \text{ milligrams}}$

  $q(8) \approx \boxed{12.5 \text{ milligrams}}$

19. a. $\dfrac{dN}{dt} = 0.0049 N(37 - N)$ countries per year

  b. This differential equation has a logistic function as its solution.

  $N(t) = \dfrac{37}{1 + Ae^{-0.0049(37)t}}$

  $\boxed{N(t) = \dfrac{37}{1 + Ae^{-0.1813t}} \text{ countries}}$

c. Solving $N(55) = 16$ yields $A \approx 28{,}097.439$

$$N(t) = \frac{37}{1 + 28{,}097.439 e^{-0.1813t}} \text{ countries}$$

d. 1840: $N(40) \approx \boxed{2 \text{ countries}}$

1860: $N(60) \approx \boxed{24 \text{ countries}}$

21. i. a.

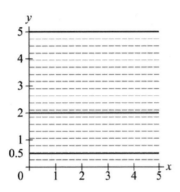

One possible answer: The displayed particular solutions go through the points $(0, 0.5)$, $(2, 2)$ and $(4, 5)$.

b. All particular solutions are horizontal lines, each passing through the chosen initial condition.

c. $y = C$, where C is a constant

ii. a.

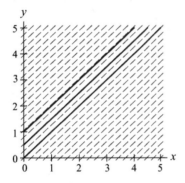

One possible answer: The displayed particular solutions go through the points $(0, 0.5)$, $(2, 2)$ and $(4, 5)$.

b. All particular solutions are lines with slope 1, each passing through the chosen initial condition.

c. $y = x + C$, where C is a constant

iii. a.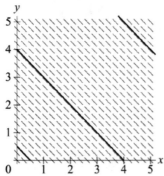

One possible answer: The displayed particular solutions go through the points (0, 0.5), (2, 2) and (4, 5).
b. All particular solutions are parallel lines with slope –1 and differing vertical shifts.
c. $y = -x + C$, where C is a constant

iv. a.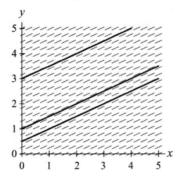

One possible answer: The displayed particular solutions go through the points (0, 0.5), (2, 2) and (4, 5).
b. All particular solutions are lines with slope $\frac{1}{2}$, each passing through the chosen initial condition.
c. $y = \frac{1}{2}x + C$, where C is a constant

23. a. i.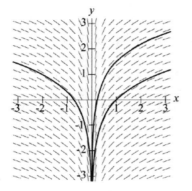

One possible answer: The particular solutions shown go through (1, 1.5), (2, 1), and (1, 0).
ii. When x > 0, the graph of a particular solution rises as x gets larger. When x < 0, the solution graph rises as x gets smaller. The particular solution graphs are concave down.

iii. The family of solutions appears to increase rapidly as x moves away from the origin (in both directions), and then the increase slows down. The line x = 0 (lying on the y-axis) appears to be a vertical asymptote for the family.

b. i.
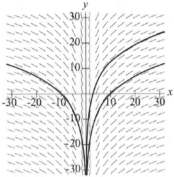
One possible answer: The particular solutions shown go through (10, 0), (̃10, 0), and (5, 5).

ii. When x > 0, the graph of a particular solution rises as x gets larger. When x < 0, the solution graph rises as x gets smaller. The particular solution graphs are concave down.

iii. The family of solutions appears to behave the same as that in part a, but the slope at each point on a particular solution graph is 10 times the slope at the corresponding point on a particular solution graph in part a. Again, the line x = 0 appears to be a vertical asymptote for the family.

c. i.
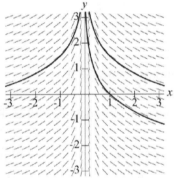
One possible answer: The particular solutions shown go through (1, 1.5), (̃2, 1), and (1, 0).

ii. When x > 0, the graph of a particular solution falls as x gets larger. When x < 0, the solution graph falls as x gets smaller. The particular solution graphs are concave up.

iii. The slope at each point on a particular solution graph is the negative of the slope at a corresponding point on a particular solution graph in part a. The family of solutions appears to decrease rapidly as x moves away from the origin (in both directions), and then the decrease levels off. The line x = 0 appears to be a vertical asymptote for the family.

d. i.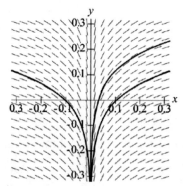

One possible answer: The particular solutions shown go through (0.1, 0), (0.1, 0), and (0.05, 0.05).

ii. When $x > 0$, the graph of a particular solution rises as x gets larger. When $x < 0$, the solution graph rises as x gets smaller. The particular solution graphs are concave down.

iii. The family of solutions appears to behave the same as that in part a, but the slope at each point on a particular solution graph is $\frac{1}{10}$ times the slope at the corresponding point on a particular solution graph in part a. Again, the line $x = 0$ appears to be a vertical asymptote for the family.

25. Solve by separation of variables.
$$\frac{1}{y}dy = kdx$$
$$\int \frac{1}{y}dy = \int kdx$$
$$\ln|y| + c_1 = kx + c_2$$
$$\ln|y| = kx + C$$
$$y = \pm e^{kx+C}$$
$$y(x) = \pm ae^{kx}$$

27. Solve by antidifferentiation.
$$\int dy = \int \frac{k}{x}dx$$
$$y(x) = k\ln|x| + C$$

29. Solve by separation of variables.

$$\frac{1}{y}dy = \frac{k}{x}dx$$

$$\int \frac{1}{y}dy = \int \frac{k}{x}dx$$

$$\ln|y| + c_1 = k\ln|x| + c_2$$

$$\ln|y| = k\ln|x| + C$$

$$|y| = e^C e^{\ln|x|^k}$$

$$|y| = a|x|^k$$

$$y(x) = \pm ax^k$$

(Note that $k\ln|x| + C$ can be rewritten as $\ln|x|^k + C$. Now change the equation from logarithmic form to exponential form. $|y| = e^{k\ln|x|+C} = e^{k\ln|x|} \cdot e^C$. Replace e^C with a and remove the absolute value signs.)

31. a. $\dfrac{df}{dx} = k$

b. $f(x) = kx + C$

c. Taking the derivative in part b, we get $\dfrac{d}{dx}(kx + C) = k$. Thus we have the identity $\dfrac{k}{x} = \dfrac{k}{x}$, and the solution is verified.

33. $\dfrac{d^2 S}{dt^2} = \dfrac{k}{S^2}$

35. $\dfrac{d^2 P}{dy^2} = k$; Taking the antiderivative, we get $\dfrac{dP}{dy} = ky + C$.

Taking the antiderivative of $\dfrac{dP}{dy}$, we get $P(y) = \dfrac{k}{2}y^2 + Cy + D$.

37. a. $\dfrac{d^2 R}{dt^2} = 6.14$ jobs per month per month in the tth month of the year

b. Taking the antiderivative of $\dfrac{d^2 R}{dt^2}$, we get $\dfrac{dR}{dt} = 6.14t + C$.

When $t = 1$, $\dfrac{dR}{dt} = -0.87$. Solving for C, we get $-0.87 = 6.14(1) + C$

$C = -7.01$

$\dfrac{dR}{dt} = 6.14t - 7.01$ jobs per month in the tth month of the year

Taking the antiderivative of $\dfrac{dR}{dt}$, we get $R(t) = 3.07t^2 - 7.01t + C$.

When $t = 2$, $R = 14$. Solving for C, we get $14 = 3.07(2^2) - 7.01(2) + C$

$C = 15.74$

The particular solution is $R(t) = 3.07t^2 - 7.01t + 15.74$ jobs in the tth month of the year

c. $R(8) \approx 156$ and $R(11) \approx 310$

We estimate the number of jobs in August to be approximately 156 and the number in November to be 310.

39. a. $\dfrac{d^2A}{dt^2} = -2009$ cases per year per year, where t is the number of years since 1988

b. Taking the antiderivative of $\dfrac{d^2A}{dt^2}$, we get $\dfrac{dA}{dt} = -2099t + C$. When $t = 0$, $\dfrac{dA}{dt} = 5988.7$, so $C \approx 5988.7$. The particular solution is $\dfrac{dA}{dt} = -2099t + 5988.7$ cases per year, where t is the number of years since 1988.

Taking the antiderivative of $\dfrac{dA}{dt}$, we get $A(t) = -1049.5t^2 + 5988.7t + C$. When $t = 0$, $A = 33{,}590$, so $C \approx 33{,}590$. The particular solution is
$A(t) = -1049.5t^2 + 5988.7t + 33{,}590$ cases, where t is the number of years since 1988

c. When $t = 3$, $\dfrac{dA}{dt} = -308.3$ and $A(3) \approx 42{,}111$.

We estimate that in 1991 there were 42,111 AIDS cases and the number of cases was decreasing at rate of 308.3 cases per year.

41. a. $\dfrac{d^2x}{dt^2} = -\dfrac{k}{m}x$, where $m = \dfrac{30}{32} = \dfrac{15}{16}$, so $\dfrac{k}{m} = \dfrac{15}{15/16} = 16$ and $\dfrac{d^2x}{dt^2} = -16x$.

b. $x(t) = a\sin(4t + c)$ and $x'(t) = 4a\cos(4t + c)$

Because $x(0) = 2$ and $x'(0) = 0$, we have $2 = a\sin c$ and $0 = 4a\cos c$. The second equation is true if $\cos c = 0$. This occurs at $c = \dfrac{\pi}{2}$. Substituting this in to the first equation, we have

$2 = a\sin\dfrac{\pi}{2} = a(1) = a$. Thus $a = 2$ and $c = \dfrac{\pi}{2}$. The particular solution is

$x(t) = 2\sin\left(4t + \dfrac{\pi}{2}\right)$ feet beyond its equilibrium point after t seconds.

c.
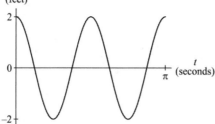

The graph shows that over time the spring oscillates from 2 feet beyond its equilibrium point and back again.

d. $x(t) = 0$ when $4t + \dfrac{\pi}{2} = \pi$ Solving for t yields $t = \dfrac{\pi}{8}$

$x'\left(\dfrac{\pi}{8}\right) = 8\cos(\dfrac{\pi}{2} + \dfrac{\pi}{2}) = 8(-1) = -8$ feet per second

The mass is moving at a speed of 8 ft/sec when it passes its equilibrium point.

43. a. $\dfrac{dT}{dt} = k(T - A)$ °F per minute

b. Solving $1.8 = k(98 - 70)$ yields $k \approx \boxed{0.064}$

$\dfrac{dT}{dt} \approx 0.064(T - 70)$ °F per minute

c. Initial condition (0, 98); $\Delta x = 1$ minute

t	Estimate of $T(t)$	$\dfrac{dT}{dt}$
0	98	−4.5
1	96.875	−4.436
2	95.766	−4.371
3	94.673	−4.307
4	93.596	−4.243
5	92.536	−4.179
6	91.491	−4.114
7	90.463	−4.050
8	89.450	−3.986
9	88.454	−3.921
10	87.473	−3.857
11	86.509	−3.793
12	85.561	−3.729
13	84.629	−3.664
14	83.713	−3.6
15	$\boxed{82.813°F}$	

Chapter 6 Review Activities

1. a. $\int_0^5 -0.0158(0.992127535^x)\,dx \approx -0.077 \;\rightarrow\; \boxed{0.077 \text{ grams}}$ decays

(0.5 watts per gram) $\cdot$ (2 − 0.077 grams) $\approx \boxed{0.962 \text{ watts}}$

b. $\int_0^\infty -0.0158(0.992127535^x)dx = \lim_{N \to \infty} \int_0^N -0.0158(0.992127535^x)dx$

$$= \lim_{N \to \infty} \frac{-0.0158(0.992127535^x)}{\ln 0.992127535}\bigg|_0^N$$

$$\approx \lim_{N \to \infty}(1.999(0.992127535^N - 1))$$

$$\approx -1.999 \quad \rightarrow \quad \boxed{1.999 \text{ grams}} \text{ decays}$$

3. a. i. $R_a(t) = (0.07)(37)$ million dollars per year

   ii. $\int_0^9 R_a(t)e^{0.045(5-t)}dt \approx \boxed{\$28.395 \text{ million}}$

   b. i. $R_b(t) = (0.07)(37 + 0.9t)$ million dollars per year

   ii. $\int_0^9 R_b(t)e^{0.045(5-t)}dt \approx \boxed{\$31.306 \text{ million}}$

   c. i. $R_c(t) = (0.07)(37)(1.032^t)$ million dollars per year

   ii. $\int_0^9 R_c(t)e^{0.045(5-t)}dt \approx \boxed{\$32.534 \text{ million}}$

5. a. survival: $1526(0.92^{25}) \approx 189.780 \quad \rightarrow \quad \boxed{190 \text{ bears}}$

   b. (The activity should give the birth rate for cubs as $r(t) = 0.180 + 0.012t$ thousand cubs per year. With a first year survival rate of 43%.)

   yearly renewal and survival for year t: $\boxed{f(t) = 0.43r(t)(0.92^{25-t})}$

   c. $189.780 + 1000\int_0^{25} f(t)dt \approx 1860.056 \quad \rightarrow \quad \boxed{1860 \text{ bears}}$

7. a. $D(p) = -50p + 124.5$ thousand bags gives the demand for cat treats in a certain region at a price of p dollars per bag, data from $0.79 to $1.89.

   b. Solving $D(p) = 0$ yields $\boxed{p = \$2.49}$

   c. $D(1.49) \cdot 1.49 + \int_{1.49}^{2.49} D(p)dp = \boxed{99.5 \text{ thousand bags}}$

   d. $\eta = \frac{p \cdot D'(p)}{D(p)} = \frac{-50p}{-50p + 124.5}$

   Solving $\eta = -1$ yields $p = 1.245$

   $D(1.245) = 62.25$

   $\boxed{(\$1.245, 62.25 \text{ thousand bags})}$

   Checking elasticity for prices on either side of $1.245 shows:

   for $p < \$1.245$, $-1 < \eta < 1 \quad \rightarrow \quad$ demand is $\boxed{\text{inelastic}}$ for prices less than $1.245

   for $p > \$1.245$, $\eta < -1 \quad \rightarrow \quad$ demand is $\boxed{\text{elastic}}$ for prices greater than $1.245

9. a. $S(p) \approx \begin{cases} 0 & \text{when } p < 1 \\ 9.134p + 7.894 & \text{when } p \geq 1 \end{cases}$ gives supply for yogurt when the price is p dollars per carton, data from $1.00 to $4.48. The output of the supply function is measured in thousand containers.

b. Solving $S(p) = 26$ yields $p \approx$ $1.98 per container

c. $2.98 \cdot S(2.98) \approx$ $104.637 thousand

d. $\int_1^{2.98} S(p)dp \approx$ $51.620 thousand

11. (In the activity the demand function is missing a negative sign before the first term. The equation should be $D(p) = -0.05p^2 - 0.82p + 6.448$.)

a. $S(3) =$ 4.737 thousand bags
$D(3) = 3.538$
Supply exceeds demand at a price of $3 per bag.

b. Solving $S(p) = D(p)$ yields $p \approx 2.105$
$S(2.105) \approx 4.500$
($2.11, 4.500 thousand bags)

c. $\int_{1.5}^{2.105} S(p)dp + \int_{2.105}^{5.807} D(p)dp \approx$ $11.445 thousand

13. a. $P(6 < x < 9) = ?$
$P(6 < x < 9) =$ 0.1576

b. $P(12 < x < 15) =$ 0.1766
Among families with annual incomes between $56,870 and $98,470, 17.66% of a family's child-care expenses was spent on a child between the ages of 12 and 15.

15. a. $P(7 < x < 8) = \int_7^8 \frac{1}{0.5\sqrt{2\pi}} e^{\frac{-(x-7.5)^2}{2(0.5)^2}} dx$

b. $P(7 < x < 8) \approx 0.68$
68% of mature golden eagles have a wingspan between 7 and 8 feet.

c. $P(x > 8.1) = \int_{8.1}^{\infty} \frac{1}{0.5\sqrt{2\pi}} e^{\frac{-(x-7.5)^2}{2(0.5)^2}} dx$

Estimate the integral numerically:

$N \to \infty$	$\int_{8.1}^{N} \frac{1}{0.5\sqrt{2\pi}} e^{\frac{-(x-7.5)^2}{2(0.5)^2}} dx$
50	0.115
100	0.115
200	0.115

$P(x > 8.1) = \lim_{N \to \infty} \int_{8.1}^{N} \frac{1}{0.5\sqrt{2\pi}} e^{\frac{-(x-7.5)^2}{2(0.5)^2}} dx \approx 0.115$

11.5% of mature golden eagles have a wingspan of more than 8.1 feet.

Chapter 7 Ingredients of Multivariable Change: Models, Graphs, Rates

Section 7.1 Multivariable Functions and Contour Graphs (pages 532–539)

1. a.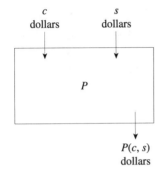

   b. i. $P(1.2, s)$ dollars is the profit from the sale of one yard of fabric with production cost of $1.20 per yard and selling price of s dollars per yard.

   ii. $P(c, 4.5)$ dollars is the profit from the sale of one yard of fabric with a production cost of c dollars per yard and a selling price of $4.50 per yard.

   iii. The profit from the sale of one yard of fabric with a production cost of $1.20 per yard and a selling price of $4.50 per yard is $3.00.

3. a.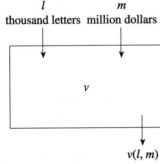

   b. i. $v(100, m)$ is the probability that a certain senator votes in favor of a bill when he receives 100,000 letters supporting the bill and m million dollars is invested in lobbying against the bill.

   ii. $v(l, 53)$ is the probability that a certain senator votes in favor of a bill when he receives l hundred thousand letters supporting the bill and $53,000,000 is invested in lobbying against the bill.

   iii. When a certain senator receives 100,000 letters supporting a bill and $53,000,000 is invested in lobbying against the bill, there is a 50% chance that the senator will support the bill.

5. a.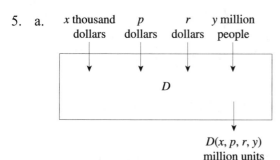

b. The demand for a consumer commodity is $D(53.7, 29.99, 154.99, 2.5)$ million units when the average household income is $53,700, the price of the commodity is $29.99, the price of a related commodity is $154.99, and the consumer base is 2,500,000 people.

c. $D(p, r)$ is the demand for a consumer commodity where the average household income is $53,700 and the consumer base is 2,500,000 people.

7. a.

```
    m        r percentage    n times      t
  dollars       points      per year    years
     │            │             │         │
     ▼            ▼             ▼         ▼
  ┌─────────────────────────────────────────┐
  │                   P                     │
  └─────────────────────────────────────────┘
                       │
                       ▼
                  P(m, r, n, t)
                     dollars
```

b. The face value of a loan is $P(500, 0.06, 12, 15)$ dollars when the payment amount is $500 paid 12 times a year for 15 years at 6% interest.

c. $P(m, t)$ dollars is the face value of a loan for which the payment amount is m dollars paid 12 times a year for t years at 6% interest.

9. (The answers in the answer key are incorrect.)

a. The position of a point on a 3D graph can be estimated by following the grid-lines to the front (or side) edge of the surface, then dropping vertically down to the axis. Estimating the input value from the axis can then be done as with two-dimensional graphs.

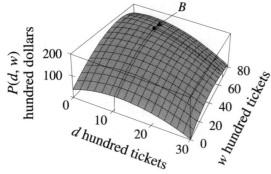

$d \approx \boxed{11 \text{ hundred tickets}}$

$w \approx \boxed{65 \text{ hundred tickets}}$

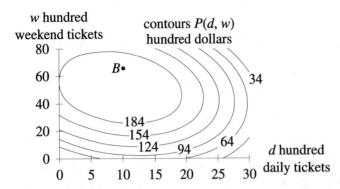

a. Because point B is not between two distinct contour curves on the contour graph, information from the 3D graph will also be used to estimate the output level of P at B. The 3D graph shows the maximum level of P to be no more than 200. Assuming 200 is the maximum output of P and estimating its position on the contour graph as the center of the 184-level contour curve, establishes a point of comparison so that the output level of P at B can be estimated.

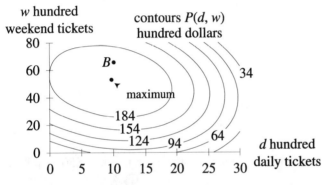

$P(11, 65) \approx \boxed{190 \text{ hundred dollars}}$

c. When 1,100 daily admission tickets and 6,500 weekend admission tickets are purchased to a weekend craft fair, the craft fair organizers realize a profit of $19,000.

11. (The answers in the answer key are incorrect.)
    a. input values: $h \approx \boxed{10 \text{ hours}}$
    $$p \approx \boxed{70\%}$$
    output value: $D(10, 70) \approx \boxed{12.6 \text{ days}}$

    C. grandis will develop in 12.6 days when there is 10 hours of light per day and 70% humidity.
    b. Input values of point B can be estimated from the 3D graph.

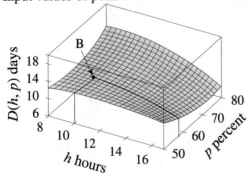

input values: $h \approx \boxed{10 \text{ hours}}$
$p \approx \boxed{57\%}$

output value: Transferring point B to the contour graph shows $D(10, 57) \approx \boxed{11.9 \text{ days}}$

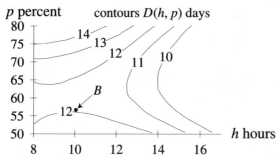

C. grandis will develop in 11.9 days when there is 10 hours of light per day and 57% humidity.

13. a.

Air Temperature (°F)	Relative humidity (%)												
	40	45	50	55	60	65	70	75	80	85	90	95	100
110	135												
108	130	137											
106	124	130	137										
104	119	124	130	137									
102	114	119	124	130	137								
100	109	113	118	123	129	136							
98	105	108	113	117	122	128	134						
96	101	104	107	111	116	121	126	132					
94	97	100	103	106	110	114	119	124	129	135			
92	94	96	98	101	104	108	112	116	121	126	131		
90	91	92	94	97	99	102	106	109	113	117	122	126	131	130
88	88	89	91	93	95	97	100	103	106	109	113	117	121
86	85	86	88	89	91	93	95	97	99	102	105	108	111
84	83	84	85	86	87	89	90	92	94	96	98	100	102	105
82	81	82	83	83	84	85	86	87	88	90	91	93	94
80	80	80	81	81	82	82	83	83	84	85	85	86	87	90

b.

Air Temperature (°F)	Relative humidity (%)												
	40	45	50	55	60	65	70	75	80	85	90	95	100
110	135												
108	130	137											
106	124	130	137										
104	119	124	130	137									
102	114	119	124	130	137								
100	109	113	118	123	129	136							
98	105	108	113	117	122	128	134						
96	101	104	107	111	116	121	126	132					
94	97	100	103	106	110	114	119	124	129	135			
92	94	96	98	101	104	108	112	116	121	126	131		
90	91	92	94	97	99	102	106	109	113	117	122	126	131	130
88	88	89	91	93	95	97	100	103	106	109	113	117	121
86	85	86	88	89	91	93	95	97	99	102	105	108	111
84	83	84	85	86	87	89	90	92	94	96	98	100	102	105
82	81	82	83	83	84	85	86	87	88	90	91	93	94
80	80	80	81	81	82	82	83	83	84	85	85	86	87	90

15. a. 12.0 hours
    b. 12.0 hours
    a. Answers will vary based on location of the school.
    b.

	Month												
North ▸	Jan	Feb	Mar	Apr	May	Jun	Jul	Aug	Sep	Oct	Nov	Dec	
South ▸	Jul	Aug	Sep	Oct	Nov	Dec	Jan	Feb	Mar	Apr	May	Jun	
0	12.1	12.1	12.1	12.1	12.1	12.1	12.1	12.1	12.1	12.1	12.1	12.1	
5	11.9	11.9	12.1	12.2	12.3	12.4	12.4	12.3	12.2	12.0	11.9	11.8	
10	11.6	11.8	12.1	12.3	12.6	12.7	12.6	12.4	12.2	11.9	11.7	11.5	
15	11.3	11.7	12.0	12.5	12.8	13.0	12.9	12.6	12.2	11.8	11.4	11.2	
20	11.1	11.5	12.0	12.6	13.1	13.3	13.2	12.8	12.3	11.7	11.2	10.9	
25	10.8	11.3	12.0	12.7	13.3	13.7	13.5	13.0	12.3	11.6	10.9	10.6	
30	10.4	11.1	12.0	12.9	13.6	14.1	13.9	13.2	12.3	11.5	10.7	10.3	
35	10.1	10.9	11.9	13.1	14.0	14.5	14.3	13.5	12.4	11.3	10.4	9.9	
40	9.7	10.7	11.9	13.2	14.4	15.0	14.7	13.8	12.5	11.2	10.0	9.4	
45	9.2	10.4	11.9	13.5	14.8	15.6	15.2	14.1	12.5	11.0	9.6	8.8	
50	8.6	10.1	11.8	13.8	15.4	16.3	15.9	14.5	12.7	10.8	9.1	8.2	
55	7.8	7.7	11.8	14.1	16.1	17.3	16.8	15.0	12.8	10.5	8.4	7.3	
60	6.8	7.2	11.7	14.6	17.1	18.7	18.0	15.7	12.9	10.2	7.6	6.0	

 Latitude (°N or °S)

17. a.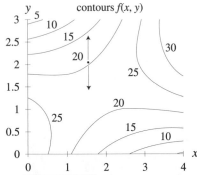

As y decreases from 2 to 1.5 along the line $x = 1.5$, the output of f increases from less than 20 to near 22. As y increases from 2, the output of f decreases.

b.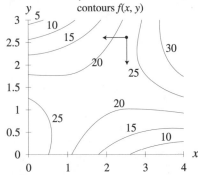

As x decreases, the output of f decreases more rapidly than it does as y decreases because the contour curves are more closely spaced to the left of (2.5, 2.5) than they are directly below that point,

c.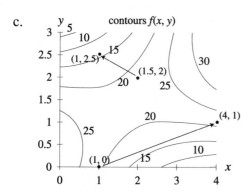

The change is greater when input shifts from (2, 2) to (1, 2.5), causing the output of f to change by approximately $21 - 15.75 \approx 5.25$, than it is when (1, 0) shifts to (4, 1), causing virtually no change in contour values.

19. a.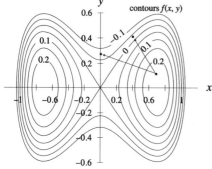

Output values of f can be estimated from the contour graph.
$f(0.4, 0.4) \approx -0.025$
$f(0, 0.3) \approx -0.075$
From $f(0.7, 0.1)$ the descent to $f(0, 0.3)$ is greater than the descent to $f(0.4, 0.4)$.

b.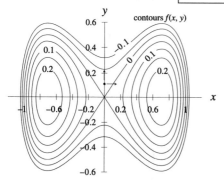

The output of f increases as x increases from 0.1 along the $y = 0$ line until $x \approx 0.7$. The output of f decreases as y increases from 0 along the $x = 0.1$ line.

c.

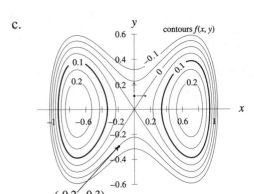

$f(-0.2, -0.3) \approx -0.05$ can be estimated from the contour graph.

Any point lying on the 0.1 contour can be used to answer this question. There are infinitely many such points. Two possibilities are (-0.7, 0.38) and (0.94, 0).

21. a.

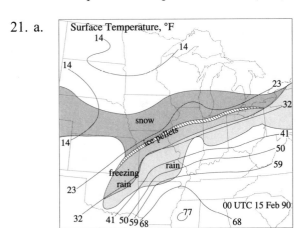

b.

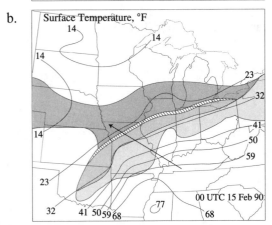

Temperature can be estimated from the contour graph.
    southern Kentucky: at least 59°F
    northern Missouri: no more than 23°F (more likely, closer to 21 or 22°F)
The temperature is $\boxed{\text{at least 36°F}}$ less in northern Missouri than it is in southern Kentucky.

23. Solving $g = 100k(1.09^s)$ for k in terms of s and g yields $k = \dfrac{g}{100(1.09^s)}$.

200-level contour: $k = \dfrac{200}{100(1.09^s)}$. Alternate form: $k = 2(1.09^{-s})$

400-level contour: $k = \dfrac{400}{100(1.09^s)}$. Alternate form: $k = 4(1.09^{-s})$

600-level contour: $k = \dfrac{600}{100(1.09^s)}$. Alternate form: $k = 6(1.09^{-s})$

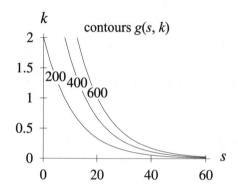

25. Solving $P = 38.6s - 2su + 13u + 0.99u^2$ for s in terms of u and P:

$$38.6s - 2su = P - 13u - 0.99u^2$$
$$s(38.6 - 2u) = P - 13u - 0.99u^2$$
$$s = \dfrac{P - 13u - 0.99u^2}{38.6 - 2u}$$

40-level contour: $s = \dfrac{40 - 13u - 0.99u^2}{38.6 - 2u}$.

60-level contour: $s = \dfrac{60 - 13u - 0.99u^2}{38.6 - 2u}$.

100-level contour: $s = \dfrac{100 - 13u - 0.99u^2}{38.6 - 2u}$.

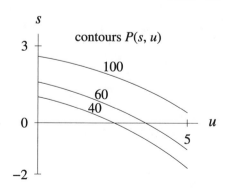

27. Solving $0.0041d^2h^{1.4} = 117$ for h in terms of d yields $h = \left(\dfrac{117}{0.0041d^2}\right)^{\frac{1}{1.4}}$

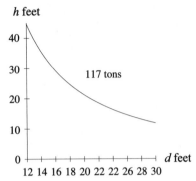

29. a. $P(c, 100, 10, s) = 109.168 - 0.730s + 0.027s^2 + 0.002(100) - 0.041(10) + 0.175c$

$\boxed{P(c, s) = 0.027s^2 - 0.730s + 0.175c + 108.958}$

b. Solving $110 = 0.027s^2 - 0.730s + 0.175c + 108.958$ for c in terms of s:

$0.027s^2 - 0.730s + 0.175c + 108.958 = 110$

$0.175c = 110 - 0.027s^2 + 0.730s - 108.958$

$c = \dfrac{1.042 - 0.027s^2 + 0.730s}{0.175}$

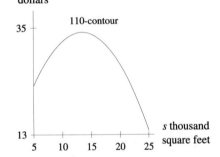

31. a. table estimate:
    133 pounds can be estimated using the 130 pound row.
    5'7" = 67 inches is half way between the table columns for 66 inches and 68 inches.
    From the table $B(66, 130) = 21.0$ and $B(68, 130) = 19.8$.
    Half-way between these two values is $\dfrac{21.0 + 19.8}{2} = \boxed{20.4}$

    function calculation: $B(67, 133) = 703\dfrac{133}{67^2} \approx \boxed{20.828}$

    BMI is in the $\boxed{\text{normal}}$ range

b.

Weight (pounds)	Height (inches)						
	60	62	64	66	68	70	72
90	17.6	16.5	15.4	14.5	13.7	12.9	12.2
100	19.5	18.3	17.2	16.1	15.2	14.3	13.6
110	21.5	20.1	18.9	17.8	16.7	15.8	14.9
120	23.4	21.9	20.6	19.4	18.2	17.2	16.3
130	25.4	23.8	22.3	21.0	19.8	18.7	17.6
140	27.3	25.6	24.0	22.6	21.3	20.1	19.0
150	29.3	27.4	25.7	24.2	22.8	21.5	20.3
160	31.2	29.3	27.5	25.8	24.3	23.0	21.7
170	33.2	31.1	29.2	27.4	25.8	24.4	23.1
180	35.2	32.9	30.9	29.0	27.4	25.8	24.4
190	37.1	34.7	32.6	30.7	28.9	27.3	25.8
200	39.1	36.6	34.3	32.3	30.4	28.7	27.1

c. Solving $K = 703\dfrac{w}{h^2}$ for w in terms of h and K yields

$$\boxed{w = \dfrac{Kh^2}{703}}$$ pounds, where h is height in inches and K is BMI

d.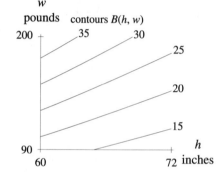

33. a. Solving $p(w, s) = K$ for s in terms of w and K:

$$10.65 + 1.13w + 1.04s - 5.83ws = K$$
$$s(1.04 - 5.83w) = K - 10.65 - 1.13w$$
$$\boxed{s = \dfrac{K - 10.65 - 1.13w}{1.04 - 5.83w}}$$

a.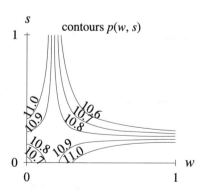

35. Answers will vary but might be similar to the following:
    A path of steepest descent is a curve crossing multiple contours in such a way that at any point on the descent path the distance along the path between that point and the next contour level is as short as possible.
    To sketch a path of steepest descent on a contour graph: locate the starting point, find the closest point on the next lower contour curve, cross this contour curve with a perpendicular line and repeat until there are no lower contours. The sketch may produce a straight line, but more often it will produce a curve. The path of steepest descent will meet all intersecting contour levels at right angles.

Section 7.2   Cross-Sectional Models and Rates of Change (pages 547–550)

1. a. "as a function of the dew point" implies that "dew point" is the variable of change.
      $\boxed{\text{Air temperature}}$ is held constant.
   b. 95°F air temperature is represented by a $\boxed{\text{row}}$.
   c. modeling the 95°F air temperature row:
      $A(p, 95) \approx 0.024p^2 - 2.345p + 151.311$ °F gives the apparent temperature when the dew point is p°F and the air temperature is 95°F, data from $50 \leq p \leq 85$.

3. a.

Fraction of sky covered	Midn.	3 A.M	6 A.M	9 A.M	Noon	3 P.M	6 P.M	9 P.M
1.0	0.45	0.48	0.47	0.45	0.41	0.39	0.41	0.38
≥0.9	0.48	0.52	0.49	0.57	0.55	0.53	0.46	0.46
≥0.8	0.50	0.55	0.52	0.62	0.60	0.57	0.49	0.43
≥0.7	0.52	0.56	0.54	0.64	0.63	0.59	0.52	0.47
≥0.6	0.54	0.59	0.57	0.66	0.65	0.61	0.54	0.49
≥0.5	0.55	0.59	0.59	0.69	0.67	0.64	0.55	0.51
≥0.4	0.56	0.59	0.62	0.72	0.68	0.66	0.58	0.56
≥0.3	0.58	0.62	0.64	0.74	0.71	0.69	0.60	0.60
≥0.2	0.60	0.67	0.66	0.78	0.74	0.71	0.62	0.62
≥0.1	0.66	0.71	0.72	0.89	0.86	0.84	0.79	0.66

Time of day

$0.60 \rightarrow \boxed{60\%}$

   b. "model … cloud cover … at 9 A.M." implies "time of day" is held constant.
      $\boxed{\text{Fraction of sky covered}}$ is the input variable of change.
   c. 9 A.M. is represented by a $\boxed{\text{column}}$.
   d. modeling the 9 A.M. column:
      $c(9, f) \approx -1.651f^3 + 2.686f^2 - 1.597f + 1.019$ gives the frequency of cloud cover at 9 A.M. (expressed as a decimal) where f is the fraction (expressed as a decimal) of the sky covered by clouds, data from $0.1 \leq f \leq 1.0$.

5. a. "for a 52-month loan" implies the loan period is held constant at 52.
      Because the table does not include a 52-month loan column, it is not possible to use this table to model monthly payments for a 52-month loan.
   b. "for a loan at 9%" implies the interest rate is held constant at 9.
      Because the table includes a row for 9% monthly interest, it is possible to use that row to model monthly payments for different terms at 9% interest.
   c. The monthly payments for a 52-month loan can be extrapolated from a model for 9% interest.
      modeling the 9% monthly interest row:

$m(t, 9) \approx 0.017t^2 - 2.091t + 85.977$ dollars gives the monthly payment on a $1000 loan when t months is the length of the loan and 9% is the interest rate, data from $24 \leq t \leq 60$.
$m(52, 9) \approx \boxed{\$22.65}$

7. a. modeling the row for $40,000 family income (the row labeled 4):
   $c(p, 4) \approx 0.893p^2 - 1.304p + 7.811$ pounds gives the per capita consumption of peaches by people living in households with $40,000 yearly income where $p + 1.50$ dollars per pound is the price of peaches, $0.00 \leq p \leq 0.50$.
   b. $1.55 per pound → $p = 1.55 - 1.50 = 0.05$
   $c(0.05, 4) \approx \boxed{7.7 \text{ pounds}}$

(The rows and columns in the table for Activities 9 and 10 should be labeled as "Price of …" instead of "Cost of…".)

9. a. The table includes a column for $1.00 Coke products but not a row for $0.90 Pepsi products. Daily sales of Coke products when the price of Pepsi products is $0.90 can be interpolated from a model of $1.00 Coke products.
   The $\boxed{\text{price of Coke products}}$ is held constant.
   b. $1.00 Coke products is represented by a $\boxed{\text{column}}$.
   c. modeling the column for $1.00 Coke products:
   $\boxed{S(1.00, p) = 196p + 25 \text{ cans gives the daily sales of Coke products when Pepsi products are sold for } \$p \text{ per can and Coke products are sold for } \$1.00 \text{ per can}, 0.50 \leq p \leq 1.50.}$
   $S(1.00, 0.90) = 201.4$ → $\boxed{201 \text{ cans}}$

11. a.

Age (years)	Year						
	1990	1995	2000	2005	2010	2015	2020
15	3.34	3.65	3.87	4.24	4.31	4.22	4.26
20	4.04	3.51	3.88	4.10	4.48	4.55	4.45
25	4.06	3.79	3.39	3.73	3.94	4.30	4.36
30	4.50	4.38	3.92	3.52	3.86	4.08	4.44
35	4.27	4.59	4.47	4.00	3.61	3.95	4.17
40	3.80	4.28	4.65	4.54	4.07	3.68	4.02
45	2.90	3.70	4.21	4.57	4.46	4.01	3.62
50	2.43	2.93	3.69	4.19	4.55	4.44	4.00

$\boxed{3.88 \text{ million people}}$

b.

Age (years)	Year						
	1990	1995	2000	2005	2010	2015	2020
15	3.34	3.65	3.87	4.24	4.31	4.22	4.26
20	4.04	3.51	3.88	4.10	4.48	4.55	4.45
25	4.06	3.79	3.39	3.73	3.94	4.30	4.36
30	4.50	4.38	3.92	3.52	3.86	4.08	4.44
35	4.27	4.59	4.47	4.00	3.61	3.95	4.17
40	3.80	4.28	4.65	4.54	4.07	3.68	4.02
45	2.90	3.70	4.21	4.57	4.46	4.01	3.62
50	2.43	2.93	3.69	4.19	4.55	4.44	4.00

c. "with respect to age" implies "age" is the variable of change.
   modeling the 2005 column:
   $p(2005, a) \approx 0.480\sin(0.197a - 0.725) + 4.060$ million people gives the population of age a years in 2005, data from $15 \leq a \leq 50$.

$$\frac{dp(2005, a)}{da} \approx 0.480\cos(0.197a - 0.725) \cdot 0.197 \quad \text{by the Chain Rule}$$

$$\left.\frac{dp(2005, a)}{da}\right|_{a=35} \approx \boxed{0.094 \text{ million people per year}}$$

13. a.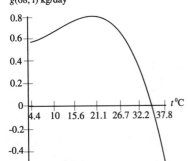

    $\boxed{-0.35 \text{ kg/day}}$

   b. (The model in the answer key is incorrect.)
      modeling the 68 kg column:
      $g(68, t) = -9.022 \cdot 10^{-5}t^3 + 0.003t^2 - 0.013t + 0.572$ kg/day gives the average daily weight gain/loss for a pig weighing 68 kg when the air temperature is $t\,°C$, data from $4.4 \le t \le 37.8$.

   c. $g(68, t)$ kg/day

   The graph of $g(68, t)$ is concave up, increasing between $t = 4.4$ and $t \approx 12.1$. After which the graph is concave down. It reaches a maximum near $t \approx 22.1$. A pig weighing 68 kilograms will gain the most weight when the temperature is near 22.1°C. For temperatures in excess of 22.1°C, the weight gain diminishes more and more rapidly until weight is actually lost at temperatures in excess of approximately 37.1°C. A pig's weight gain is greatest when the air temperature is moderate.

15. a. modeling the $14,000 column:
       $F(14,000, r) = 1.4r^2 + 280r + 14,000$ dollars gives the value of $14,000 after two years invested at $r\%$ interest compounded annually, data from $2 \le r \le 12$.

    b. $\dfrac{dF(14,000, r)}{dr} = 2.8r + 280$

       $\left.\dfrac{dF(14,000, r)}{dr}\right|_{r=12.7} \approx \boxed{\$315.56 \text{ per percentage point}}$

    c. $F(14,000, r) = 14,000r^2 + 28,000r + 14,000$ dollars gives the value of $14,000 after two years invested at $100r\%$ interest compounded annually, data from $0.02 \le r \le 0.12$
       The coefficient of r^2 is multiplied by a factor of 10,000 (or 100^2). The coefficient of r is multiplied by a factor of 100. The constant term remains the same.

d. $\dfrac{dF(14,000, r)}{dr} = 28,000r + 28,000$ dollars per 100 percentage points

$\left.\dfrac{dF(14,000, r)}{dr}\right|_{r=0.127} \approx \boxed{\$31{,}556 \text{ per 100 percentage points}}$

The derivative tells us approximately how much the output will change when the input increases by one unit. If the input is a percentage expressed as a decimal, then an increase in one unit corresponds to 100 percentage points. For example, if $r = 0.127$ and is increased by 1 to $r = 1.127$ the corresponding percentages are 12.7% and 112.7%.
This answer is equivalent to the one found in part b.

17. Answers will vary but might be similar to the following:
Derivatives of cross-sectional models of a function with two input variables can be used to pinpoint optimal points on those cross-sections. These optimal points in turn may be used to estimate the point at which a critical point may appear on the three-dimensional function.

Section 7.3 Partial Rates of Change (pages 557–561)

1. $\dfrac{\partial w}{\partial h}$ pounds per inch

3. $\left.\dfrac{\partial t}{\partial y}\right|_{x=23}$ degrees per degree (degrees Fahrenheit per degree latitude)

5. $\left.\dfrac{\partial r}{\partial c}\right|_{b=2}$ dollars per cow for $c = 100$

7. a. $\left.\dfrac{\partial p}{\partial m}\right|_{l=100{,}000}$ is the rate of change of the probability that the senator will vote for the bill with respect to the amount spent on lobbying when the senator receives 100,000 letters in opposition to the bill.
If the number of letters is constant but lobbying funding against the bill increases, the probability that the senator votes for the bill may decline and the rate of change will be negative.

   b. $\left.\dfrac{\partial p}{\partial l}\right|_{m=53}$ is the rate of change of the probability that the senator will vote for the bill with respect to the number of letters received when $53 million is spent on lobbying efforts.
If the number of letters increases (while lobbying funding remains constant) the probability that the senator votes for the bill is likely to increase and the rate of change will be positive.

9. a. $f(x, y) = 3x^2 + 5xy + 2y^3$

   $\dfrac{\partial f}{\partial x} = 3 \cdot 2x + 5(1)y + 0$

   $\boxed{\dfrac{\partial f}{\partial x} = 6x + 5y}$

b. $f(x, y) = 3x^2 + 5xy + 2y^3$

$\dfrac{\partial f}{\partial y} = 0 + 5x \cdot 1 + 2 \cdot 3y^2$

$\boxed{\dfrac{\partial f}{\partial y} = 5x + 6y^2}$

c. $\dfrac{\partial f}{\partial x}\bigg|_{y=7} = 6x + 5(7) = \boxed{6x + 35}$

11. a. $f(x, y) = 5x^3 + 3x^2 y^3 + 9xy + 14x + 8$

$\dfrac{\partial f}{\partial x} = 5 \cdot 3x^2 + 3(2x)y^3 + 9(1)y + 14 \cdot 1 + 0$

$\boxed{\dfrac{\partial f}{\partial x} = 15x^2 + 6xy^3 + 9y + 14}$

b. $f(x, y) = 5x^3 + 3x^2 y^3 + 9xy + 14x + 8$

$\dfrac{\partial f}{\partial y} = 0 + 3x^2 \cdot 3y^2 + 9x \cdot 1 + 0 + 0$

$\boxed{\dfrac{\partial f}{\partial y} = 9x^2 y^2 + 9x}$

c. $\dfrac{\partial f}{\partial x}\bigg|_{y=2} = 15x^2 + 6x(2^3) + 9(2) + 14 = \boxed{15x^2 + 48x + 32}$

13. a. $m(t, s) = s \ln t + 3.75s + 14.96$

$m_t = s \cdot \dfrac{1}{t} + 0 + 0$ by the Natural Log and Sum Rules

$\boxed{m_t = \dfrac{s}{t}}$

b. $m(t, s) = s \ln t + 3.75s + 14.96$

$m_s = 1 \cdot \ln t + 3.75 \cdot 1 + 0$

$\boxed{m_s = \ln t + 3.75}$

c. $m_s\big|_{t=3} = \boxed{\ln 3 + 3.75}$

15. a. $h(s, t, r) = \dfrac{s}{t} + \dfrac{t}{r} - (st - tr)^2$

$\dfrac{\partial h}{\partial s} = \dfrac{1}{t} + 0 - 2(st - tr) \cdot t$ by the Constant Multiplier, Difference, and Chain Rules

$\boxed{\dfrac{\partial h}{\partial s} = \dfrac{1}{t} - 2t(st - tr)}$

7.3 Partial Rates of Change
Solutions to Odd-Numbered Activities

b. Rewrite $h(s, t, r) = \dfrac{s}{t} + \dfrac{t}{r} - (st-tr)^2$ as $h(s, t, r) = st^{-1} + \dfrac{t}{r} - (st-tr)^2$

$\dfrac{\partial h}{\partial t} = s(-1t^{-2}) + \dfrac{1}{r} - 2(st-tr)(s-r)$ by the Power, Sum, and Chain Rules

$\boxed{\dfrac{\partial h}{\partial t} = \dfrac{-s}{t^2} + \dfrac{1}{r} - 2(st-tr)(s-r)}$

c. Rewrite $h(s, t, r) = \dfrac{s}{t} + \dfrac{t}{r} - (st-tr)^2$ as $h(s, t, r) = \dfrac{s}{t} + tr^{-1} - (st-tr)^2$

$\dfrac{\partial h}{\partial r} = 0 + t(-r^{-2}) + 2t(st-tr)$ by the Power, Sum, and Chain Rules

$\boxed{\dfrac{\partial h}{\partial r} = \dfrac{-t}{r^2} + 2t(st-tr)}$

d. $\left.\dfrac{\partial h}{\partial r}\right|_{(s,\,t,\,r)=(1,\,2,\,-1)} = \dfrac{-2}{(-1)^2} + 2(2)(1(2) - 2(-1)) = \boxed{14}$

17. $f(x, y) = 2xy + 8x^2 y^3 + 5e^{2y} + 10$

f_x: $f(\mathbf{x}, y) = 2\mathbf{x}y + 8\mathbf{x}^2 y^3 + 5e^{2y} + 10$

       $f_x = 2(1)y + 8(2x)y^3 + 0 + 0$

       $\boxed{f_x = 2y + 16xy^3}$

f_{xx}: $f_x = 2y + 16\mathbf{x}y^3$

       $f_{xx} = 0 + 16(1)y^3$

       $\boxed{f_{xx} = 16y^3}$

f_{xy}: $f_x = 2\mathbf{y} + 16x\mathbf{y}^3$

       $f_{xy} = 2(1) + 16x(3y^2)$

       $\boxed{f_{xy} = 2 + 48xy^2}$

f_y: $f(x, \mathbf{y}) = 2x\mathbf{y} + 8x^2 \mathbf{y}^3 + 5e^{2\mathbf{y}} + 10$

       $f_y = 2x + 8x^2(3y^2) + 5(2e^{2y}) + 0$

       $\boxed{f_y = 2x + 24x^2 y^2 + 10e^{2y}}$

f_{yx}: $f_y = 2\mathbf{x} + 24\mathbf{x}^2 y^2 + 10e^{2y}$

       $f_{yx} = 2 + 24(2x)y^2 + 0$

       $\boxed{f_{yx} = 2 + 48xy^2}$ $f_{yx} = f_{xy}$ even though they are written in a different order

f_{yy}: $f_y = 2x + 24x^2 \mathbf{y}^2 + 10e^{2\mathbf{y}}$

       $f_{yy} = 0 + 24x^2(2y) + 10(2e^{2y})$

       $\boxed{f_{yy} = 48x^2 y + 20e^{2y}}$

19. $f(x, y) = \dfrac{x}{y} - \dfrac{y}{x}$

f_x: Rewrite $f(x, y) = \dfrac{x}{y} - \dfrac{y}{x}$ as $f(x, y) = xy^{-1} - x^{-1}y$

$f_x = y^{-1} - (-1x^{-2})y$

$\boxed{f_x = y^{-1} + x^{-2}y}$ Alternate form: $f_x = \dfrac{1}{y} + \dfrac{y}{x^2}$

f_{xx}: $f_x = y^{-1} + x^{-2}y$

$f_{xx} = 0 + (-2x^{-3})y$

$\boxed{f_{xx} = -2x^{-3}y}$ Alternate form: $f_{xx} = \dfrac{-2y}{x^3}$

f_{xy}: $f_x = y^{-1} + x^{-2}y$

$f_{xy} = -y^{-2} + x^{-2} \cdot 1$

$\boxed{f_{xy} = -y^{-2} + x^{-2}}$ Alternate form: $f_{xy} = -\dfrac{1}{y^2} + \dfrac{1}{x^2}$

f_y: $f(x, y) = xy^{-1} - x^{-1}y$

$f_y = x(-1y^{-2}) - x^{-1} \cdot 1$

$\boxed{f_y = -xy^{-2} - x^{-1}}$ Alternate form: $f_y = \dfrac{-x}{y^2} - \dfrac{1}{x}$

f_{yx}: $f_y = -xy^{-2} - x^{-1}$

$f_{yx} = -(1)y^{-2} - (-x^{-2})$

$\boxed{f_{yx} = -y^{-2} + x^{-2}}$ $f_{yx} = f_{xy}$

f_{yy}: $f_y = -xy^{-2} - x^{-1}$

$f_{yy} = -x(-2y^{-3}) - 0$

$\boxed{f_{yy} = 2xy^{-3}}$ Alternate form: $f_{yy} = \dfrac{2x}{y^3}$

21. $h(x, y) = e^{2x-3y}$ The Product Rule is used for each of the partials and second partials.

h_x: $h(x, y) = e^{2x-3y}$

$h_x = (2x - 3y)e^{2x-3y} \cdot (2 + 0)$

$\boxed{h_x = 2(2x - 3y)e^{2x-3y}}$

h_{xx}: $h_x = 2(2x - 3y)e^{2x-3y}$

$h_{xx} = 2(2x - 3y) \cdot (2x - 3y)e^{2x-3y} \cdot (2 + 0)$

$\boxed{h_{xx} = 4(2x - 3y)^2 e^{2x-3y}}$

h_{xy}: $h_x = 2(2x - 3y)e^{2x-3y}$

$$h_{xy} = 2(2x-3y)\cdot(2x-3y)e^{2x-3y}\cdot(0-3)$$
$$\boxed{h_{xy} = -6(2x-3y)^2 e^{2x-3y}}$$

h_y: $h(x, y) = e^{2x-3y}$
$$h_y = (2x-3y)e^{2x-3y}\cdot(0-3)$$
$$\boxed{h_y = -3(2x-3y)e^{2x-3y}}$$

h_{yx}: $h_{yx} = -3(2\boldsymbol{x}-3y)e^{2\boldsymbol{x}-3y}$
$$h_{xy} = -3(2x-3y)\cdot(2x-3y)e^{2x-3y}\cdot(2+0)$$
$$\boxed{h_{xy} = -6(2x-3y)^2 e^{2x-3y}} \qquad h_{xy} = h_{yx}$$

h_{yy}: $h_y = -3(2x-3y)e^{2x-3y}$
$$h_{xy} = -3(2x-3y)\cdot(2x-3y)e^{2x-3y}\cdot(0-3)$$
$$\boxed{h_{xy} = 9(2x-3y)^2 e^{2x-3y}}$$

23. a. $F(14{,}000, r) = 14{,}000(1+r)^2$ dollars gives the value of an investment after 2 years when the APY is $100r\%$.

b. $\left.\dfrac{\partial F}{\partial r}\right|_{P=14{,}000} = 28{,}000(1+r)$

$\left.\dfrac{\partial F}{\partial r}\right|_{(P, r) = (14{,}000, \, 0.1272)} \approx \boxed{\$31{,}561.60 \text{ per 100 percentage points}}$

$\boxed{\text{The future value of an investment of \$14{,}000 after 2 years in an account with an APY of 12.72\%,}}$
$\boxed{\text{is increasing by \$315.62 per percentage point.}}$

c. The slope of the line tangent to a graph of $F(14{,}000, r)$ at $r = 0.1272$ is \$31,562 per 100 percentage points.

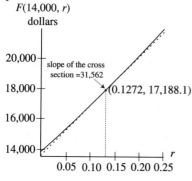

25. a. "with respect to yearly income" implies i is the variable of change.
$$C(p, \boldsymbol{i}) = 2\ln\boldsymbol{i} + 2.7183^{-p} + 4$$
$$C_i = \dfrac{2}{i} + 0 + 0$$
$\left.C_i\right|_{(p, \, i)=(0.2, \, 3)} \approx \boxed{0.67 \text{ pounds per \$10{,}000}}$

b. "with respect to price" implies p is the variable of change.

$C(p, i) = 2\ln i + 2.7183^{-p} + 4$

$C_p = 0 + \ln(2.7183)(-2.7183^{-p}) + 0$

$C_p\big|_{(p, i)=(0.2, 3)} \approx \boxed{-0.819 \text{ pounds per dollar/pound}}$

27. a. "with respect to air temperature" implies t is the variable of change:

$H(v, t) = (10.45 + 10\sqrt{v} - v) \cdot (33 - t)$

When v is treated as a constant, the factor $(10.45 + 10\sqrt{v} - v)$ can be considered to be a constant multiplier:

$H_t = (10.45 + 10\sqrt{v} - v) \cdot \dfrac{\partial}{\partial t}(33 - t)$

$= (10.45 + 10\sqrt{v} - v) \cdot (0 - 1)$

$\boxed{H_t = -10.45 - 10\sqrt{v} + v \text{ kg-calories per °C}}$

"with respect to wind speed" implies v is the variable of change:

Rewrite H as $H(v, t) = (10.45 + 10v^{0.5} - v) \cdot (33 - t)$

When t is treated as a constant, the factor $(33 - t)$ can be considered to be a constant multiplier:

$H_v = \left(\dfrac{\partial}{\partial v}(10.45 + 10v^{0.5} - v)\right) \cdot (33 - t)$

$= (0 + 5v^{-0.5} - 1) \cdot (33 - t)$

$\boxed{H_v = (5v^{-0.5} - 1)(33 - t) \text{ kg-calories per mps}}$

Alternate form: $H_v = (\dfrac{5}{\sqrt{v}} - 1)(33 - t)$

b. $\dfrac{\partial H}{\partial v}$ should be positive because an increase in wind speed (when temperature is constant) should cause an increase heat loss.

$\dfrac{\partial H}{\partial v}\bigg|_{(v, t)=(20, 12)} \approx \boxed{2.48 \text{ kg-calories per mps}}$

c. $\dfrac{\partial H}{\partial t}$ should be negative because an increase in temperature (when wind speed is constant) should cause a decrease heat loss.

$\dfrac{\partial H}{\partial t}\bigg|_{(v, t)=(20, 12)} \approx \boxed{-35.17 \text{ kg-calories per degree Celsius}}$

7.3 Partial Rates of Change
Solutions to Odd-Numbered Activities

29. a. Food intake should increase as either milk production or size increases.

b. $\dfrac{\partial I}{\partial s}$: $I(s,m) = 8.62 - 1.24s + 0.09s^2 - 0.21m + 0.036m^2 + 0.21sm$

$\dfrac{\partial I}{\partial s} = 0 - 1.24 + 0.18s + 0 + 0 + 0.21m$

$\boxed{\dfrac{\partial I}{\partial s} = -1.24 + 0.18s + 0.21m \text{ kilograms per unit change}}$

(kilograms organic matter per unit increase in the size index)

$\dfrac{\partial I}{\partial s}$: $I(s,m) = 8.62 - 1.24s + 0.09s^2 - 0.21m + 0.036m^2 + 0.21sm$

$\dfrac{\partial I}{\partial m} = 0 + 0 + 0 - 0.21 + 0.072m + 0.21s$

$\boxed{\dfrac{\partial I}{\partial m} = -0.21 + 0.072m + 0.21s \text{ kilograms per kilogram}}$

(kilograms organic matter per kilograms milk)

c. "milk production increases" implies m is the variable of change.

$\left.\dfrac{\partial I}{\partial m}\right|_{(s,m)=(2,6)} = \boxed{0.642 \text{ kilograms per kilogram}}$

d. "size increases" implies s is the variable of change.

$\left.\dfrac{\partial I}{\partial s}\right|_{(s,m)=(2,6)} = \boxed{0.38 \text{ kilograms per unit change}}$

31. a. $\dfrac{\partial F}{\partial t}$: $F(t, r) = 1000e^{rt}$

$\boxed{\dfrac{\partial F}{\partial t} = 1000re^{rt} \text{ dollars per year}}$

$\dfrac{\partial F}{\partial r}$: $F(t, r) = 1000e^{rt}$

$\boxed{\dfrac{\partial F}{\partial r} = 1000te^{rt} \text{ dollars per 100 percentage points}}$

b. $\dfrac{\partial F}{\partial t}$: $\dfrac{\partial F}{\partial t} = 1000re^{rt}$ dollars per year

$\dfrac{\partial^2 F}{\partial t^2}$: $\dfrac{\partial F}{\partial t} = 1000re^{rt}$

$\boxed{\dfrac{\partial^2 F}{\partial t^2} = 1000r(re^{rt})}$ (The Product Rule is not necessary for this derivative.)

Alternate form: $\dfrac{\partial^2 F}{\partial t^2} = 1000r^2 e^{rt}$

$$\left.\frac{\partial^2 F}{\partial t^2}\right|_{(30, \, 0.047)} \approx 9.05$$

The rate of change of the value of a $1000 investment earning 4.7% compounded continuously for 30 years is increasing by $9.05 per year per year.

$\dfrac{\partial^2 F}{\partial t \partial r}$: $\dfrac{\partial F}{\partial t} = 1000re^{rt}$

$\dfrac{\partial^2 F}{\partial t \partial r} = \left(\dfrac{\partial}{\partial r} 1000r\right) \cdot e^{rt} + 1000r \cdot \dfrac{\partial}{\partial r} e^{rt}$ by the Product Rule

$= 1000 \cdot e^{rt} + 1000r \cdot te^{rt}$

$$\boxed{\dfrac{\partial^2 F}{\partial t \partial r} = 1000e^{rt} + 1000rte^{rt}}$$

Alternate form: $\dfrac{\partial^2 F}{\partial t \partial r} = 1000(e^{rt} + rte^{rt})$

$$\left.\dfrac{\partial^2 F}{\partial t \partial r}\right|_{(30, \, 0.047)} \approx 9871.25$$

The rate of change of the value of a $1000 investment earning 4.7% compounded continuously for 30 years is increasing by approximately $98.71 per year per percentage point.

$\dfrac{\partial F}{\partial r}$: $\dfrac{\partial F}{\partial r} = 1000te^{rt}$ dollars per 100 percentage points

$\dfrac{\partial^2 F}{\partial r \partial t}$: $\dfrac{\partial F}{\partial r} = 1000te^{rt}$

$\dfrac{\partial^2 F}{\partial r \partial t} = \left(\dfrac{\partial F}{\partial t} 1000t\right) \cdot e^{rt} + 1000t \cdot \dfrac{\partial F}{\partial t} e^{rt}$ by the Product Rule

$= 1000 \cdot e^{rt} + 1000t \cdot re^{rt}$

$$\boxed{\dfrac{\partial^2 F}{\partial r \partial t} = 1000e^{rt} + 1000rte^{rt}} \qquad \dfrac{\partial^2 F}{\partial r \partial t} = \dfrac{\partial^2 F}{\partial t \partial r}$$

Alternate form: $\dfrac{\partial^2 F}{\partial r \partial t} = 1000(e^{rt} + rte^{rt})$

$$\left.\dfrac{\partial^2 F}{\partial r \partial t}\right|_{(30, \, 0.047)} \approx 9871.25$$

The rate of change of the value of a $1000 investment earning 4.7% compounded continuously for 30 years is increasing by $98.71 per percentage point year.

$\dfrac{\partial^2 F}{\partial r^2}$: $\dfrac{\partial F}{\partial r} = 1000te^{rt}$

$\dfrac{\partial^2 F}{\partial r^2} = \left(\dfrac{\partial F}{\partial r} 1000t\right) \cdot e^{rt} + 1000t \cdot \dfrac{\partial F}{\partial r} e^{rt}$

$= 0 \cdot e^{rt} + 1000t \cdot te^{rt}$

$$\boxed{\dfrac{\partial^2 F}{\partial r^2} = 1000t(te^{rt})}$$

Alternate form: $\dfrac{\partial^2 F}{\partial r^2} = 1000 t^2 e^{rt}$

$\left.\dfrac{\partial^2 F}{\partial r^2}\right|_{(30,\ 0.047)} \approx 3{,}686{,}359.86$

The rate of change of the value of a $1000 investment earning 4.7% compounded continuously for 30 years is increasing $368.636 per percentage point per percentage point.

33. a. $F(t, r) = (1+r)^t$

$\dfrac{\partial F}{\partial t} = \ln(1+r)(1+r)^t$ million dollars per year   by the Exponential Rule $\left(\dfrac{d}{dt} b^t = \ln b (b^t)\right)$

b. $F(t, r) = (1+r)^t$

$\dfrac{\partial F}{\partial r} = t(1+r)^{t-1}$ million dollars per 100 percentage points

c. "with respect to time" implies t is the variable of change.

$\left.\dfrac{\partial F}{\partial t}\right|_{(0.15,\ 5)} \approx 0.28$ million dollars per year

d. (The figure in the answer key has the wrong variable on the vertical axis.)

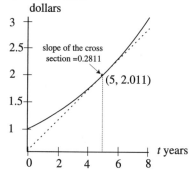

35. Answers will vary but should be similar to the following:
According to the constant multiplier rule, any constant factor in the function will also be a factor of the derivative. So, it doesn't matter whether the constant is substituted before or after applying the derivative.

Section 7.4 Compensating for Change (pages 567–569)

1. $f(x, y) = 15x^2 y^3$
   $f_x = 30xy^3$
   $f_y = 45x^2 y^2$
   $\dfrac{dx}{dy} = \dfrac{-f_y}{f_x}$

   $$\boxed{\dfrac{dx}{dy} = \dfrac{-45x^2 y^2}{30xy^3}}$$

   Alternate form: $\dfrac{dx}{dy} = \dfrac{-3x}{2y}$

3. $g(m, n) = 59.3 \ln m + 49mn + 16$
   $g_m = \dfrac{59.3}{m} + 49n$
   $g_n = 49m$
   $\dfrac{dm}{dn} = \dfrac{-g_n}{g_m}$

   $$\boxed{\dfrac{dm}{dn} = \dfrac{-49m}{\dfrac{59.3}{m} + 49n}}$$

   Alternate form: $\dfrac{dm}{dn} = \dfrac{-49m^2}{59.3 + 49mn}$

For Activities 5 through 8, even though contours and tangent lines could be drawn with either input variable on the horizontal axis, we present the contours with the first input variable on the horizontal axis and the second input variable on the vertical axis.

5. $g(x, y) = x(1.05^y)$
   Graph the contour by solving for an equation for the specific contour curve:
   Solving $x(1.05^y) = 100$ for y in terms of x:
   $$1.05^y = \dfrac{100}{x}$$
   $$\ln 1.05^y = \ln\left(\dfrac{100}{x}\right)$$
   $$y \ln 1.05 = \ln\left(\dfrac{100}{x}\right)$$
   $$y = \ln\left(\dfrac{100}{x}\right) \div \ln 1.05$$

Solving $g(x, 5) = 100$ yields $x \approx 78.353$

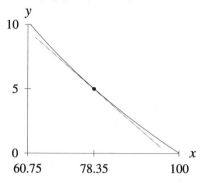

Calculating the slope of the tangent line:
$$g_x = 1.05^y$$
$$g_y = x(\ln 1.05)1.05^y$$
$$\frac{dy}{dx} = \frac{-g_x}{g_y}$$
$$= \frac{-1.05^y}{x(\ln 1.05)1.05^y}$$
$$= \frac{-1}{x(\ln 1.05)}$$
$$\left.\frac{dy}{dx}\right|_{(78.353, 5)} \approx \boxed{-0.262}$$

7. $f(a, b) = 2.8a^2b^3 - 1.8a^2 + 12b$

   Graph the contour by plotting points (three possible points are calculated):
   Solving $f(0, b) = 15$ yields $b = 1.25$
   Solving $f(a, 0.9) = 15$ yields $a \approx 4.173$
   Solving $f(10, b) = 15$ yields $b \approx 0.870$

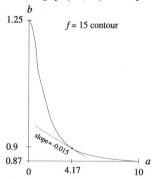

Calculating the slope of the tangent line:
$$f_a = 5.6ab^3 - 3.6a$$
$$f_b = 8.4a^2b^2 + 12$$

$$\frac{db}{da} = \frac{-f_a}{f_b}$$

$$= \frac{-(5.6ab^3 - 3.6a)}{8.4a^2b^2 + 12}$$

$$\left.\frac{db}{da}\right|_{(4.173,\ 0.9)} \approx \boxed{-0.015}$$

9. a. $f(2,\ 1) = \boxed{21}$
    b. $f_m = 6m + 2n$
        $f_n = 2m + 10n$
        $$\frac{dn}{dm} = \frac{-f_m}{f_n}$$
        $$= \frac{-6m + 2n}{2m + 10n}$$
        $$\left.\frac{dn}{dm}\right|_{(2,\ 1)} = \frac{-14}{14} = -1$$
        $\Delta m = 0.2$
        $$\Delta n \approx \frac{dn}{dm}\Delta m$$
        $$= (-1)(0.2)$$
        $$= \boxed{-0.2}$$

11. (The answers in the answer key are incorrect.)
    a. $f(3.5,\ 1148) \approx 3.703$
    b. $f_h = 0.00091s[0.103(\ln 2.5)(2.5^h)]$
        $f_s = 0.00091[0.103(2.5^h) + 1]$
        $$\frac{ds}{dh} = \frac{-f_h}{f_s}$$
        $$\left.\frac{ds}{dh}\right|_{(3.5,\ 1148)} \approx -755.144$$
        $\Delta h = -0.5$
        $$\Delta s \approx \frac{ds}{dh}\Delta h$$
        $$\approx (-755.144)(-0.5)$$
        $$\approx \boxed{377.57}$$

13. a. $A(6, 250) \approx \boxed{\$7.162}$

   b. "when more T-shirts are printed" implies n is the variable of change.
   $$\frac{\partial A}{\partial n} = (-0.02c^2 + 0.35c + 0.99)(\ln 0.99897)(0.99897^n)$$
   $$\left.\frac{\partial A}{\partial n}\right|_{(6, 250)} \approx \boxed{-0.002 \text{ dollars per shirt}}$$

   c. $A_c = (-0.04c + 0.35)(0.99897^n) + 0.46$
   $$A_n = \frac{\partial A}{\partial n} = (-0.02c^2 + 0.35c + 0.99)(\ln 0.99897)(0.99897^n)$$
   $$\frac{dn}{dc} = \frac{-A_c}{A_n}$$
   $$\boxed{\frac{dn}{dc} = \frac{-[(-0.04c + 0.35)(0.99897^n) + 0.46]}{(-0.02c^2 + 0.35c + 0.99)(\ln 0.99897)(0.99897^n)} \text{ shirts per color}}$$

   $\boxed{\text{If the number of colors increases, the order size would also need to increase to keep average cost constant so } \frac{dn}{dc} \text{ should be positive.}}$

   d. $\Delta n \approx \left.\frac{dn}{dc}\right|_{(4, 500)} \Delta c$
   $= (450)(1)$
   $= \boxed{450 \text{ shirts per color}}$

15. a. Graph the contour $p(r, t) = 53$ by plotting points (four possible points are calculated):
   Solving $p(20, t) = 53$ yields $t \approx 87.204$
   (It also yields $t \approx 43.828$ which is outside of the specified range for t.)
   Solving $p(44, t) = 53$ yields $t \approx 86.513$
   (It also yields $t \approx 44.518$ which is outside of the specified range for t.)
   Solving $p(r, 86.5) = 53$ yields $r \approx 23.125$ and $r \approx 43.934$
   (Both solutions are within the specified input interval.)

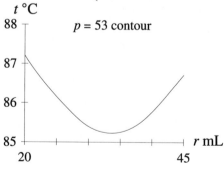

b. The points of tangency for this part were calculated in the solution to part a as (23.125, 86.5) and (43.934, 86.5).

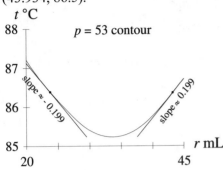

c. $p_t = -9.6544 + 0.14736t$
$p_r = 1.9836 - 0.05916r$

$$\frac{dt}{dr} = \frac{-p_r}{p_t}$$

$$\boxed{\frac{dt}{dr} = \frac{-(1.9836 - 0.05916r)}{-9.6544 + 0.14736t}} \text{ °C per milliliter}$$

d. slope at (23.125, 86.5): $\left.\frac{dt}{dr}\right|_{23.125,\ 86.5} \approx \boxed{-0.199 \text{ °C per milliliter}}$

slope at (43.934, 86.5): $\left.\frac{dt}{dr}\right|_{43.934,\ 86.5} \approx \boxed{0.199 \text{ °C per milliliter}}$.

17. a. $B(67, 129) \approx 20.202$

Solving $703\frac{w}{h^2} = 20.202$ for w yields $\boxed{w = \frac{20.202h^2}{703}}$ pounds

b. $\frac{dw}{dh} = \frac{2(20.202)h}{703}$

$\left.\frac{dw}{dh}\right|_{(67,\ 129)} \approx \boxed{3.851 \text{ pounds per inch}}$

19. a. $m(10{,}000, 5) \approx \boxed{\$193.31}$

b. $m_A = \frac{0.005}{1 - 0.9419^t}$

$m_t = \frac{0.005A}{(1 - 0.9419^t)^2}(\ln 0.9419)0.9419^t$

$\Delta A \approx \left.\frac{dA}{dt}\right|_{(10{,}000,\ 5)} \Delta t \approx -\$1715.63$

amount that could be borrowed $\approx \$10{,}000 - \$1715.63 = \boxed{\$8284.37}$

Chapter 7 Review Activities (pages 570–573)

1. a. $R(c, 13)$ billion dollars gives the revenue from transporting 13,000 passengers and c thousand tons of cargo on international flights.
    b. $R(689, p)$ billion dollars gives the revenue from transporting 689 thousand tons of cargo and p thousand passengers on international flights.
    c. The transportation of 863 thousand tons of cargo and 7,000 passengers generates $624 billion in revenue.
    d.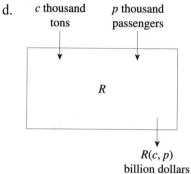

3. a.

	Width (feet)						
Length (feet)	1	2	3	4	5	6	8
1	12	12	12	—	—	—	—
2	12	18	18	12	—	—	—
3	—	12	18	18	12	—	—
4	—	—	12	18	18	12	—
5	—	—	12	18	24	18	12
6	—	12	18	24	24	24	12
7	12	18	24	24	24	24	12
8	12	18	24	24	24	18	12
9	12	12	18	18	18	12	—
10	—	—	12	12	12	—	—

    b.

	Width (feet)						
Length (feet)	1	2	3	4	5	6	8
1	12	12	12	—	—	—	—
2	12	18	18	12	—	—	—
3	—	12	18	18	12	—	—
4	—	—	12	18	18	12	—
5	—	—	12	18	24	18	12
6	—	12	18	24	24	24	12
7	12	18	24	24	24	24	12
8	12	18	24	24	24	18	12
9	—	12	18	18	18	12	—
10	—	—	12	12	12	—	—

5. (The estimates in the answer key are incorrect.)
   a. A: The input values for point A can be estimated from the 3D graph.

   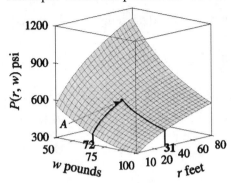

   $w \approx \boxed{72 \text{ pounds}}$

   $r \approx \boxed{31 \text{ feet}}$

   The output value for point A can be estimated by transferring the input coordinates to the contour graph and estimating the contour level.

   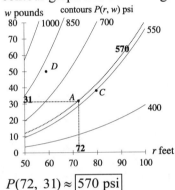

   $P(72, 31) \approx \boxed{570 \text{ psi}}$

   B: (There is no point B given in the activity)

   The input values as well as the output values for points C and D can be estimated from the contour graph.

   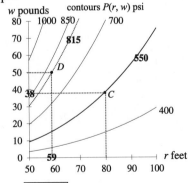

   C: $r \approx \boxed{80 \text{ feet}}$

   $w \approx \boxed{38 \text{ pounds}}$

   $P(80, 38) \approx \boxed{550 \text{ psi}}$

D: $r \approx \boxed{59 \text{ feet}}$

$w \approx \boxed{50 \text{ pounds}}$

$P(59, 50) \approx \boxed{815 \text{ psi}}$

b. $P(r, 50) \approx \dfrac{47{,}892.409}{r}$ psi gives the pressure felt by a diver when an explosion with force equivalent to 50 pounds of TNT occurs underwater r feet away from the diver.

$P(100, w) = 130\sqrt[3]{w}$ psi gives the pressure felt by a diver when an explosion with force equivalent to w pounds of TNT occurs 100 feet away from the diver.

c.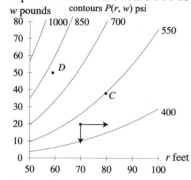

When the starting point is (70, 20), a 10-pound decrease in w results in a point closer to the 400-level contour then does a 10-foot increase in r. P decreases more quickly when $\boxed{w \text{ decreases}}$ then when r increases.

7. a. $Q(0.3, 0.4) \approx \boxed{1.058 \text{ cubic centimeters per second}}$

b. Solving $\dfrac{16.625\pi(0.22^4)}{L} = 0.35$ yields $L \approx \boxed{0.350 \text{ cm}}$

c.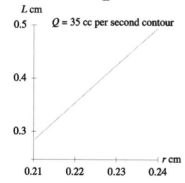

9. a. (The model in the answer key is incorrect.)
Modeling the 2005 row:
$S(q) \approx 0.475q^2 - 2.059q + 45.125$ cents per pound gives the price of sugar at the end of quarter q in 2005, data from $1 \leq q \leq 4$.

b. Modeling the 4th quarter column:
$S(t) \approx -0.078t^3 + 1.412t^2 - 5.447t + 47.410$ cents per pound gives the price of sugar t years since 2000, data from 2001 through 2009.

c. "with respect to the year" implies t is the variable of change.

Using $S(t) \approx -0.078t^3 + 1.412t^2 - 5.447t + 47.410$:

$\dfrac{dS}{dt} \approx -0.233t^2 + 2.824t - 5.447$

$S'(8) \approx \boxed{2.21 \text{ cents/pound per year}}$

d. Using $S(q) \approx 0.475q^2 - 2.059q + 45.125$:

$S'(q) \approx 0.950q - 2.059$

$S'(2) = \boxed{-0.159 \text{ cents/pound per quarter}}$

At the end of the 2nd quarter of 2005, the price of refined sugar was decreasing by 0.159 cents per pound per quarter.

11. a. "with respect to ... distance..." implies r is the variable of change.

Rewrite $P(r, w) = \dfrac{13000\sqrt[3]{w}}{r}$ as $P(r, w) = 13000r^{-1} \cdot \sqrt[3]{w}$

$\boxed{\dfrac{\partial P}{\partial r} = 13000(-r^{-2}) \cdot \sqrt[3]{w} \text{ psi per foot}}$

Alternate form: $\dfrac{\partial P}{\partial r} = \dfrac{-13000\sqrt[3]{w}}{r^2}$ psi per foot

b. "with respect to ... distance..." implies r is the variable of change.

$\boxed{\dfrac{\partial P}{\partial r}\bigg|_{w=50} = 13000(-r^{-2}) \cdot \sqrt[3]{50} \text{ psi per foot}}$

c. "weight ... may vary" implies w is the variable of change.

Rewrite $P(r, w) = \dfrac{13000\sqrt[3]{w}}{r}$ as $P(r, w) = \dfrac{13000w^{\frac{1}{3}}}{r}$

$\dfrac{\partial P}{\partial w} = \dfrac{13000\left(\frac{1}{3}w^{-\frac{2}{3}}\right)}{r}$

$\boxed{\dfrac{\partial P}{\partial w}\bigg|_{r=30} = \dfrac{13000\left(\frac{1}{3}w^{-\frac{2}{3}}\right)}{30} \text{ psi per pound}}$

13. $f(x, y, z) = 2x^3y^{-2} + e^{-xy} - 4x\ln z$

$\dfrac{\partial f}{\partial x}:$ $f(\boldsymbol{x}, y, z) = 2\boldsymbol{x}^3 y^{-2} + e^{-\boldsymbol{x}y} - 4\boldsymbol{x}\ln z$

$f_x = 2(3x^2)y^{-2} - ye^{-xy} - 4(1)\ln z$

$\boxed{f_x = 6x^2y^{-2} - ye^{-xy} - 4\ln z}$

$\dfrac{\partial f}{\partial y}:$ $f(x, \boldsymbol{y}, z) = 2x^3\boldsymbol{y}^{-2} + e^{-x\boldsymbol{y}} - 4x\ln z$

$f_y = 2x^3(-2y^{-3}) - xe^{-xy} + 0$

$\boxed{f_y = -4x^3y^{-3} - xe^{-xy}}$

$\dfrac{\partial f}{\partial z}$: $f(x, y, z) = 2x^3 y^{-2} + e^{-xy} - 4x \ln z$

$$f_z = 0 + 0 - 4x \cdot \dfrac{1}{z}$$

$$\boxed{f_z = \dfrac{-4x}{z}}$$

15. a. $T(300, 250) \approx \boxed{\$34{,}812.50}$

b. $T_p = -0.50p + 120 + 0.25s$

$T_s = 0.25p - 0.750s + 100$

$\dfrac{ds}{dp} = \dfrac{-T_p}{T_s}$

$= \dfrac{-(-0.50p + 120 + 0.25s)}{0.25p - 0.750s + 100}$

$\boxed{\dfrac{ds}{dp} = \dfrac{0.50p - 120 - 0.25s}{0.25p - 0.750s + 100}}$ chains per patch

c. $\left.\dfrac{ds}{dp}\right|_{(300,\,250)} = \boxed{2.6 \text{ chains per patch}}$

Chapter 8 Analyzing Multivariable Change

Section 8.1 Extreme Points and Saddle Points (pages 583–587)

1. Answers will vary but should be similar to the following:
   a. A relative maximum occurs where a table value is greater than all 8 values surrounding it.
   b. A relative minimum occurs where a table value is less than all 8 values surrounding it.
   c. If a table value is larger than the two adjacent table values in either the column or the row or a diagonal and is smaller than the two adjacent table values in another direction, then the value corresponds to a saddle point.
   d. If all the edges of a table are terminal edges, then the absolute maximum and minimum are simply the largest and smallest values in the table. If it is unknown whether all the edges are terminal edges, it is not possible to determine whether absolute extrema exist using the table.

3. The point A is a relative maximum point because A lies within concentric simple closed curves and the value of each curve is higher than the next larger curve: 290, 300, 310, 320, A. (All of the curves shown in the figure would be completed to simple closed curves if the input intervals for x and y were expanded.)

5. a.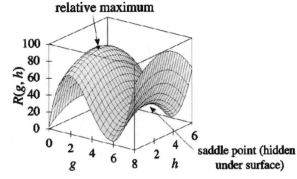

   b.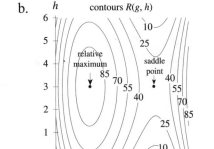

   relative maximum point: $(g, h, r) \approx (2, 3, 95)$

   saddle point: $(g, h, r) \approx (6, 3, 30)$

7. (The activity should ask about the point (1.7, 1.4) instead of (1.8, 1.5).)
   The point is a saddle point.
   The 3D graph of z appears to be shaped like a saddle with the point at (1.7, 1.4) right in the center of the saddle.

   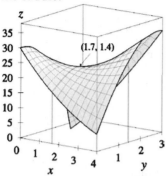

   On the contour graph of z along a line extending from (2.6, 0) through (1.7, 1.4) the output value $z(1.7, 1.4)$ is a maximum and along a line extending from (0, 0.6) to (1.7, 1.4) the output value $z(1.7, 1.4)$ is a minimum.
   (Other pairs of intersecting lines can be used with this argument.)

   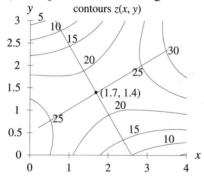

9. No edges can be considered terminal, so relative extrema do not exist along edges.
   relative maximum points:

	\multicolumn{5}{c}{Month}
Year	Feb	Mar	Apr	May	June
2004	80.5	81.3	80.1	71.0	75.1
2005	73.0	82.9	100.4	92.6	89.5
2006	79.4	81.5	86.9	96.7	84.8
2007	92.0	91.5	98.6	87.9	85.6
2008	89.5	87.3	90.2	86.8	86.0
2009	93.0	87.5	90.7	88.7	87.6

   greater than all eight adjacent table values (April, 2005, 100.4 cents per pound)

	Feb	Mar	Apr	May	June
2004	80.5	81.3	80.1	71.0	75.1
2005	73.0	82.9	100.4	92.6	89.5
2006	79.4	81.5	86.9	96.7	84.8
2007	92.0	91.5	98.6	87.9	85.6
2008	89.5	87.3	90.2	86.8	86.0
2009	93.0	87.5	90.7	88.7	87.6

   greater than all eight adjacent table values (April, 2007, 98.6 cents per pound)

relative minimum point:

Year \ Month	Feb	Mar	Apr	May	June
2004	80.5	81.3	80.1	71.0	75.1
2005	73.0	82.9	100.4	92.6	89.5
2006	79.4	81.5	86.9	96.7	84.8
2007	92.0	91.5	98.6	87.9	85.6
2008	89.5	**87.3**	90.2	86.8	86.0
2009	93.0	87.5	90.7	88.7	87.6

less than all eight adjacent table values
(March, 2008, 87.3 cents per pound)

saddle points:

Year \ Month	Feb	Mar	Apr	May	June
2004	80.5	81.3	80.1	71.0	75.1
2005	73.0	82.9	100.4	92.6	89.5
2006	79.4	81.5	86.9	96.7	84.8
2007	92.0	**91.5**	98.6	87.9	85.6
2008	89.5	87.3	90.2	86.8	86.0
2009	93.0	87.5	90.7	88.7	87.6

vertical maximum, horizontal minimum
(March, 2007, 91.5 cents per pound)

Year \ Month	Feb	Mar	Apr	May	June
2004	80.5	81.3	80.1	71.0	75.1
2005	73.0	82.9	100.4	92.6	89.5
2006	79.4	81.5	86.9	96.7	84.8
2007	92.0	91.5	98.6	87.9	85.6
2008	89.5	87.3	**90.2**	86.8	86.0
2009	93.0	87.5	90.7	88.7	87.6

vertical minimum, horizontal maximum
(April, 2008, 90.2 cents per pound)

11. a. The table is complete. It is terminal at both the north and south poles. However, the table wraps so that that the first column, Jan, could follow the last column, Dec.

   b.

   c. Because the table wraps, output values for January and December should be compared against each other.

relative maximum points:
- (June, North Pole, 8.9 kW-h)
- (June, 40° North, 9.3 kW-h)
- (December, South Pole, 9.4 kW-h)
- (December, 40° South, 9.9 kW-h)

Note: (January, 30° and 40° South, 9.6 kW-h) are not relative maximum points because they are lower than the corresponding points in December.

relative minimum points: The dashes in the table can be considered to be radiation levels that are essentially 0. Considering which columns and rows tended to give lowest values suggests that the relative minima occurred at the following two points:
- (January, North Pole, 0 kW-h)
- (June, South Pole, 0 kW-h)

saddle points:
- (April, 10° North, 8.5 kW-h)   vertical maximum, diagonal minimum
- (August, 10° North, 8.4 kW-h)   horizontal maximum, diagonal minimum
- (June, 70° North, 8.5 kW-h)   vertical minimum, horizontal maximum
- (December, 70° South, 9.1 kW-h) vertical minimum, horizontal maximum

d. absolute maximum: 9.9 kW-h at (December, 40° South)

absolute minimum: 0 kW-h at (January, North Pole) and (June, South Pole)

13. a. 1.09 kilograms occurs on a terminal edge at (15.6°C, 156 kg). This table value is greater than the relative maximum value 1.01 kilograms that occurs at (21.1°C, 91 kg).

b. The maximum average daily weight gain for pigs between the weights of 45 and 156 kilograms is 1.09 kilograms for a 156-kilogram pig at an air temperature of 15.6°C.

15. a.

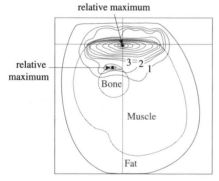

The relative maximum where fat meets muscle is 11.
The relative maximum where muscle meets bone is 4.
(The activity does not specify output units.)

b. Both points correspond to relative maximum temperature changes. The relative maximum point where fat meets muscle is also the absolute maximum point.

17. a.

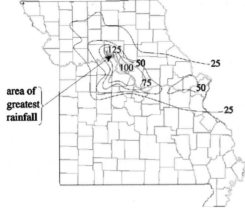

125 millimeters

b.

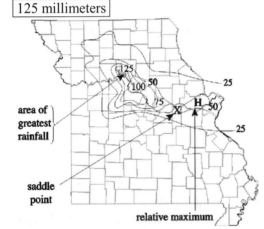

Section 8.2 Multivariable Optimization (pages 597–599)

1. $R(k, m) = 3k^2 - 2km - 20k + 3m^2 - 4m + 60$

   Setting both partial derivatives of R equal to zero gives the following system of equations:
   $$\begin{cases} R_k: & 6k - 2m - 20 = 0 \\ R_m: & -2k + 6m - 4 = 0 \end{cases}$$
   Solving the system of partial derivatives yields $m = 2$ and $k = 4$ with $R(4, 2) = 16$.
   (For algebra and matrix help see the bulleted sections following the main solution.)
   The critical point is $(4, 2, 16)$.
   Using the Determinant Test to identify the type of critical point:
   The second partial derivatives are $R_{kk} = 6$, $R_{mm} = 6$, and $R_{km} = R_{mk} = -2$
   $$D(4, 2) = \begin{vmatrix} 6 & -2 \\ -2 & 6 \end{vmatrix}$$
   $$= 6 \cdot 6 - (-2)(-2)$$
   $$= 32$$
   Because $D > 0$ and $R_{kk} > 0$, the critical point is a relative minimum point.

- Algebra help for solving the system of partial derivatives:
  Rewrite the system of partial derivatives as
  $$\begin{cases} 6k - 2m = 20 & [1] \\ -2k + 6m = 4 & [2] \end{cases}$$
  Eliminate k by multiplying equation [1] by 3 and adding the result to equation [2]:
  $$\begin{array}{r} 6k - 2m = 20 \\ (+) \; -6k + 18m = 12 \\ \hline 0 + 16m = 32 \end{array}$$
  Solving $16m = 32$ yields $m = 2$.
  Substituting $m = 2$ into equation [1] gives $6k - 2(2) = 20$.
  Solving $6k - 4 = 20$ yields $k = 4$.
  $R(4, 2) = 16$

- Alternate solution using matrices:
  Rewrite the system of partial derivatives as
  $$\begin{cases} 6k - 2m = 20 \\ -2k + 6m = 4 \end{cases}$$
  Represent the system using matrices:
  $$|A| = \begin{vmatrix} 6 & -2 \\ -2 & 6 \end{vmatrix}, \; |B| = \begin{vmatrix} 20 \\ 4 \end{vmatrix}, \text{ and } |X| = \begin{vmatrix} k \\ m \end{vmatrix}$$
  Solve as $|X| = |A|^{-1} \cdot |B| = \begin{vmatrix} 4 \\ 2 \end{vmatrix}$

3. $G(t, p) = pe^t - 3p$
   Setting both partial derivatives of G equal to zero gives the following system of equations:
   $$\begin{cases} G_t: \; pe^t = 0 & [1] \\ G_p: \; e^t - 3 = 0 & [2] \end{cases}$$
   Solving the system of partial derivatives yields $t = \ln 3$ and $p = 0$ with $G(\ln 3, 0) = 0$.
   - Algebra help for solving the system of partial derivatives:
     Solving equation [2] for t:
     $$e^t - 3 = 0$$
     $$e^t = 3$$
     $$t = \ln 3 \quad \text{by the natural log of each side}$$
     Substituting $t = \ln 3$ into equation [1] gives $pe^{\ln 3} = 0$.
     Solving $pe^{\ln 3} = 0$ yields $p = 0$.
     $G(\ln 3, 0) = 0$
   - Matrices are not used to solve this system because the partial derivatives are not linear.

   The critical point is $\boxed{(\ln 3, 0, 0)}$.

Using the Determinant Test to identify the type of critical point:
The second partial derivatives are $G_{tt} = pe^t$, $G_{pp} = 0$, and $G_{tp} = G_{pt} = e^t$

$$D(\ln 3, \ 0) = \begin{vmatrix} 0 & 3 \\ 3 & 0 \end{vmatrix} = -9$$

Because $D(\ln 3, \ 0) < 0$, the critical point is a saddle point.

5. $h(w, z) = 0.6w^2 + 1.3z^3 - 4.7wz$

Setting both partial derivatives of h equal to zero gives the following system of equations:
$$\begin{cases} h_w: \ 1.2w - 4.7z = 0 & [1] \\ h_z: \ 3.9z^2 - 4.7w = 0 & [2] \end{cases}$$

Solving the system of partial derivatives yields the two solutions:
$z = 0$ with corresponding $w = 0$ and $h(0, 0) = 0$
$z \approx 4.720$ with corresponding $w \approx 18.487$ and $h(18.487, \ 4.720) \approx -68.353$
(For algebra help see the bulleted sections following the main solution.)
The critical points are $(0, \ 0, \ 0)$ and $(18.487, \ 4.720, \ -68.353)$.

Using the Determinant Test to identify the types of points:
The second partial derivatives are: $h_{ww} = 1.2$, $h_{zz} = 7.8z$, and $h_{wz} = h_{zw} = -4.7$

$$D = \begin{vmatrix} 1.2 & -4.7 \\ -4.7 & 7.8z \end{vmatrix}$$

For the critical point $(0, 0, 0)$,
$$D(0, \ 0) = \begin{vmatrix} 1.2 & -4.7 \\ -4.7 & 0 \end{vmatrix} = -20.89$$

Because $D(0, \ 0) < 0$, $(0, 0, 0)$ is a saddle point.

For the critical point $(18.487, 4.720, -68.353)$
$$D(18.487, \ 4.720) \approx \begin{vmatrix} 1.2 & -4.7 \\ -4.7 & 36.816 \end{vmatrix} \approx 22.089$$

Because $D(18.487, \ 4.720) > 0$ and $f_{xx}(18.487, \ 4.720) > 0$, $(18.487, 4.720, -68.353)$ is a relative minimum point.

•Algebra help for solving the system of partial derivatives:
Solving equation [1] for w:
$$1.2w - 4.7z = 0$$
$$1.2w = 4.7z$$
$$w = \frac{4.7z}{1.2} \quad [3]$$

Substituting $w = \frac{4.7z}{1.2}$ into equation [2] gives
$$3.9z^2 - 4.7 \cdot \frac{4.7z}{1.2} = 0 \quad [4]$$

Solving equation [4] yields two solutions:
$z = 0$
Substituting $z = 0$ into equation [3] yields $w = 0$.
$h(0, 0) = 0$
$z \approx 4.720$
Substituting $z \approx 4.720$ into equation [3] yields $w \approx 18.487$
$h(18.487, 4.720) \approx -68.353$

• Matrices are not used to solve this system because the partial derivatives are not both linear.

7. $f(x, y) = 3x^2 - x^3 + 12y^2 - 8y^3 + 60$

Setting both partial derivatives of f equal to zero gives the following system of equations:
$$\begin{cases} f_x: & 6x - 3x^2 = 0 \quad [1] \\ f_y: & 24y - 24y^2 = 0 \quad [2] \end{cases}$$

Solving the system of partial derivatives yields four solutions:
$x = 0$ and $y = 0$ with corresponding output $f(0, 0) = 60$
$x = 0$ and $y = 1$ with corresponding output $f(0, 1) = 64$
$x = 2$ and $y = 0$ with corresponding output $f(2, 0) = 64$
$x = 2$ and $y = 1$ with corresponding output $f(2, 1) = 68$

(For algebra help see the bulleted section following the main solution.)
The four critical points are $(0, 0, 60)$, $(0, 1, 64)$, $(2, 0, 64)$, and $(2, 1, 68)$.
Using the Determinant Test to identify the types of critical points:
The second partial derivatives are $f_{xx} = 6 - 6x$, $f_{yy} = 24 - 48y$, and $f_{xy} = f_{yx} = 0$.

$$D = \begin{vmatrix} 6 - 6x & 0 \\ 0 & 24 - 48y \end{vmatrix}$$

For the critical point $(0, 0, 60)$,
$$D(0, 0) = \begin{vmatrix} 6 & 0 \\ 0 & 24 \end{vmatrix} = 144$$

Because $D(0, 0) > 0$, and $f_{xx}(0, 0) > 0$, $(0, 0, 60)$ is a relative minimum point.

For the critical point $(0, 1, 64)$,
$$D(0, 1) = \begin{vmatrix} 6 & 0 \\ 0 & -24 \end{vmatrix} = -144$$

Because $D(0, 1) < 0$, $(0, 1, 64)$ is a saddle point.

For the critical point $(2, 0, 64)$,
$$D(2, 0) = \begin{vmatrix} -6 & 0 \\ 0 & 24 \end{vmatrix} = -144$$

Because $D(2, 0) < 0$, $(2, 0, 64)$ is a saddle point.

For the critical point $(2, 1, 68)$,
$$D(2, 1) = \begin{vmatrix} -6 & 0 \\ 0 & -24 \end{vmatrix} = 144$$

Because $D(2, 1) > 0$, and $f_{xx}(2, 1) < 0$, $(2, 1, 68)$ is a relative maximum point.

- Algebra help for solving the system of partial derivatives:
  - Solving equation [1] yields two solutions:
    - $x = 0$
    - $x = 2$
  - Solving equation [2] yields two solutions:
    - $y = 0$
    - $y = 1$
  - Because solutions for x and y are found independent of each other, all four combinations of x-y pairs are critical points.
- Matrices are not used to solve this system because the partial derivatives are not both linear.

9. a. $R(b, p) = 14b - 3b^2 - bp - 2p^2 + 12p$

   Setting both partial derivatives of R equal to zero gives the following system of equations:
   $$\begin{cases} R_b: & 14 - 6b - p = 0 \quad [1] \\ R_p: & -b - 4p + 12 = 0 \quad [2] \end{cases}$$

   Solving the system of partial derivatives yields the following:
   - ground beef: $b \approx 1.913$ $\rightarrow$ $\boxed{\$1.91}$ per pound
   - sausage: $p \approx 2.522$ $\rightarrow$ $\boxed{\$2.52}$ per pound

   - Algebra help for solving the system of partial derivatives:
     - Solving equation [1] for p gives
       $$p = 14 - 6b \quad [3]$$
     - Substituting equation [3] into equation [2] gives
       $$-b - 4(14 - 6b) + 12 = 0 \quad [4]$$
     - Solving equation [4] yields $b \approx 1.913$
     - Substituting $b \approx 1.913$ into equation [3] gives $p \approx 2.522$

   - Alternate solution using matrices:
     - Rewrite the system of partial derivatives as
       $$\begin{cases} -6b - 1p = -14 \\ -1b - 4p = -12 \end{cases}$$
     - Represent the system using matrices:
       $$|A| = \begin{vmatrix} -6 & -1 \\ -1 & -4 \end{vmatrix}, \ |B| = \begin{vmatrix} -14 \\ -12 \end{vmatrix}, \ \text{and } |X| = \begin{vmatrix} b \\ p \end{vmatrix}$$
     - Solve as $|X| = |A|^{-1} \cdot |B| = \begin{vmatrix} 1.913 \\ 2.522 \end{vmatrix}$

   b. The Determinant Test is used to identify the type of critical point located in part a. The second partials R_{bb} and R_{pp} are both negative and the determinant of the second partials matrix is positive. According to the Determinant Test the critical point located in part a is a maximum.

   The second partial derivatives are $R_{bb} = -6$, $R_{pp} = -4$, and $R_{bp} = R_{pp} = -1$

   $$D = \begin{vmatrix} -6 & -1 \\ -1 & -4 \end{vmatrix} = 23$$

   Because $D(1.913, 2.522) > 0$ and $R_{bb} < 0$, the critical point is a relative maximum point.

   c. $R(1.913, 2.522) \approx 28.522$ $\rightarrow$ $\boxed{\$28.52 \text{ thousand}}$

11. a. $P(x, y) = -0.002x^2 + 20x + 12.8y - 0.05y^2$

Setting both partial derivatives of P equal to zero gives the following system of equations:
$$\begin{cases} P_x: & -0.004x + 20 = 0 \quad [1] \\ P_y: & 12.8 - 0.10y = 0 \quad [2] \end{cases}$$

Solving the system of partial derivatives yields the following:

suckers: $x = 5000$ thousand pounds

peppermint sticks: $y = 128$ thousand pounds

profit: $P(5000, 128) = \$50{,}819.2$ thousand

$\boxed{(5000 \text{ thousand pounds, } 128 \text{ thousand pounds, } 50{,}819.2 \text{ thousand dollars})}$

• Algebra help for solving the system of partial derivatives:

Solving equation [1] for x yields $x = 5000$.
Solving equation [2] for y yields $y = 128$.
$P(5000, 128) = 50{,}819.2$

• Matrices are not used to solve this system because the partial derivatives are trivial.

b. Using the Determinant Test to identify the type of critical point located in part a:

The second partial derivatives are $P_{xx} = -0.004$, $P_{yy} = -0.10$, and $P_{xy} = P_{yx} = 0$

$$D = \begin{vmatrix} -0.004 & 0 \\ 0 & -0.10 \end{vmatrix} = 0.0004$$

Because $D(5000, 128) > 0$ and $P_{xx} < 0$, $(5000, 128, 50{,}819.2)$ is a relative maximum point.

13. a. $Q(x, y) = -0.9x^3 + 38x^2 + 15y^2 - 0.1y^3$

Setting both partial derivatives of Q equal to zero gives the following system of equations:
$$\begin{cases} Q_x: & -2.7x^2 + 76x = 0 \quad [1] \\ Q_y: & 30y - 0.3y^2 = 0 \quad [2] \end{cases}$$

Solving the system of partial derivatives yields four solutions of which the following has the greatest output:

cotton: $x \approx 28.148$ thousand bales

resin: $y = 100$ hundred pounds $\rightarrow$ 10,000 pounds

fabric: $Q(28.148, 100) \approx 60{,}036.031$ million linear feet

$\boxed{(28.148 \text{ thousand bales, } 100 \text{ hundred pounds, } 60{,}036.031 \text{ million linear feet})}$

• Algebra help for solving the system of partial derivatives:

Solving equation [1] yields two solutions
$x = 0$
$x \approx 28.148$
Solving equation [2] yields
$y = 0$
$y = 100$
Because solutions for x and y are found independent of each other, all four combinations of x-y pairs are critical points.
$Q(0, 0) = 0$
$Q(0, 100) = 50{,}000$

$$\begin{vmatrix} Q(28.148, 0) \approx 10{,}036.031 \\ Q(28.148, 100) \approx 60{,}036.031 \end{vmatrix}$$

• Matrices are not used to solve this system because the partial derivatives are not linear.

b. Using the Determinant Test to identify the type of critical point found in part *a*:

The second partial derivatives are $Q_{xx} = -5.4x + 76$, $Q_{yy} = 30 - 0.6y$, and $Q_{xy} = Q_{yx} = 0$

$$D = \begin{vmatrix} -5.4x + 76 & 0 \\ 0 & 30 - 0.6y \end{vmatrix}$$

$$D(28.148, 0) \approx \begin{vmatrix} -75.999 & 0 \\ 0 & -30 \end{vmatrix} = 2279.97$$

Because $D(28.148, 0) > 0$ and $Q_{xx}(28.148, 0) < 0$, $(22.148, 100, 60{,}036.031)$ is the location of a relative maximum point..

15. (The equation in the activity is missing a *y* in the third term. The equation should be
$P(x, y) = 2.14 - 0.26x - 0.34y - 0.23x^2 - 0.16y^2 - 0.25xy$.)

a.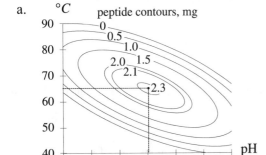

The location of the center of the concentric contour curves is estimated from the contour graph as (9, 65).

maximum peptide level: 2.35 mg
pH: 9
temperature: 65°C

b. $P(x, y) = 2.14 - 0.26x - 0.34y - 0.23x^2 - 0.16y^2 - 0.25xy$

Setting both partial derivatives of *P* equal to 0 gives the following system of equations:

$$\begin{cases} P_x: & -0.26 - 0.46x - 0.25y = 0 \quad [1] \\ P_y: & -0.34 - 0.32y - 0.25x = 0 \quad [2] \end{cases}$$

Solving the system of partial derivatives yields the following:

pH: $x \approx 0.021$ → $x + 9 \approx 9.021$

temperature: $y \approx -1.079$ → $70 + 5y \approx 64.605°C$

peptide level: 2.321 mg

(For algebra and matrix help see the bulleted sections following the main solution.)

Using the Determinant Test to identify the type of critical point:
The second partial derivatives are $P_{xx} = -0.46$, $P_{yy} = -0.32$, and $P_{xy} = P_{yx} = -0.25$

$$D = \begin{vmatrix} -0.46 & -0.25 \\ -0.25 & -0.32 \end{vmatrix} = 0.0847$$

Because $D > 0$ and $P_{xx} < 0$, the critical point is a relative maximum point.

- Algebra help for solving the system of partial derivatives:
  Solving equation [1] for y:
  $$-0.26 - 0.46x - 0.25y = 0$$
  $$0.25y = -0.26 - 0.46x$$
  $$y = \frac{-0.26 - 0.46x}{0.25}$$
  $$y = -1.04 - 1.84x \quad [3]$$
  Substituting equation [3] into equation [2] gives
  $$-0.34 - 0.32(-1.04 - 1.84x) - 0.25x = 0 \quad [4]$$
  Solving equation [4] yields $x \approx 0.021$
  Substituting $x \approx 0.021$ into equation [3] gives $y \approx -1.079$
  $P(0.021, -1.079) \approx 2.321$

- Alternate solution using matrices:
  Rewrite the system of partial derivatives as
  $$\begin{cases} -0.46x - 0.25y = 0.26 \\ -0.25x - 0.32y = 0.34 \end{cases}$$
  Represent the system using matrices:
  $$|A| = \begin{vmatrix} -0.46 & -0.25 \\ -0.25 & -0.32 \end{vmatrix}, \quad |B| = \begin{vmatrix} 0.26 \\ 0.34 \end{vmatrix}, \text{ and } |X| = \begin{vmatrix} x \\ y \end{vmatrix}$$
  Solve as $|X| = |A|^{-1} \cdot |B| = \begin{vmatrix} 0.021 \\ -1.079 \end{vmatrix}$

- Algebra help for solving the system of partial derivatives:
  Solving equation [1] for t:
  $$-0.26 - 0.46x - 0.07t = 0$$
  $$0.07t = -0.26 - 0.46x$$
  $$t = \frac{-0.26 - 0.46x}{0.07} \quad [3]$$
  Substituting equation [3] into equation [2] gives
  $$-0.04 - 0.08(\frac{-0.26 - 0.46x}{0.07}) - 0.07x = 0 \quad [4]$$
  Solving equation [4] yields $x \approx -0.564$
  Substituting $x \approx -0.564$ into equation [3] gives $t \approx -0.006$
  $P(-0.564, -0.006) \approx 2.213$

- Alternate solution using matrices:
  Rewrite the system of partial derivatives as
  $$\begin{cases} -0.46x - 0.07t = 0.26 \\ -0.07x - 0.08t = 0.04 \end{cases}$$

Represent the system using matrices:
$$|A| = \begin{vmatrix} -0.46 & -0.07 \\ -0.07 & -0.08 \end{vmatrix}, |B| = \begin{vmatrix} 0.26 \\ 0.04 \end{vmatrix}, \text{ and } |X| = \begin{vmatrix} x \\ t \end{vmatrix}$$

Solve as $|X| = |A|^{-1} \cdot |B| = \begin{vmatrix} -0.564 \\ -0.006 \end{vmatrix}$

17. (This activity should state that the constant temperature is $x\,^\circ C$ and the relative humidity is $y\%$.)

   a. $f(x,y) = -4191.6877 + 299.7038x + 23.1412y - 5.2210x^2 - 0.0937y^2 - 0.4023xy$

   Setting both partial derivatives of f equal to 0 gives the following system of equations:
   $$\begin{cases} f_x: & 299.7038 - 10.4420x - 0.4023y = 0 \quad [1] \\ f_y: & 23.1412 - 0.1874y - 0.4023x = 0 \quad [2] \end{cases}$$

   Solving the system of partial derivatives yields the critical point
   $\boxed{(26.103\%, \ 67.454\,^\circ C, \ 500.343 \text{ eggs})}$.

   (For algebra and matrix help see the bulleted sections following the main solution.)
   Using the Determinant Test to identify the type of critical point:
   The second partial derivatives are $f_{xx} = -10.4420$, $f_{yy} = -0.1874$, and $f_{xy} = f_{yx} = -0.4023$

   $$D = \begin{vmatrix} -10.4420 & -0.4023 \\ -0.4023 & -0.1874 \end{vmatrix} \approx 1.795$$

   Because $D > 0$ and $f_{xx} < 0$, the critical point is a relative maximum point.

   • Algebra help for solving the system of partial derivatives:
   Solving equation [1] for y:
   $$299.7038 - 10.4420x - 0.4023y = 0$$
   $$0.4023y = 299.7038 - 10.4420x$$
   $$y = \frac{299.7038 - 10.4420x}{0.4023} \quad [3]$$
   Substituting equation [3] into equation [2] gives
   $$23.1412 - 0.1874\left(\frac{299.7038 - 10.4420x}{0.4023}\right) - 0.4023x = 0 \quad [4]$$
   Solving equation [4] yields $x \approx 26.103$
   Substituting $x \approx 26.103$ into equation [3] gives $y \approx 67.454$
   $f(26.103, \ 67.454) \approx 500.343$

   • Alternate solution using matrices:
   Rewrite the system of partial derivatives as
   $$\begin{cases} -10.4420x - 0.4023y = -299.7038 \\ -0.4023x - 0.1874y = -23.1412 \end{cases}$$
   Represent the system using matrices:
   $$|A| = \begin{vmatrix} -10.4420 & -0.4023 \\ -0.4023 & -0.1874 \end{vmatrix}, |B| = \begin{vmatrix} -299.7038 \\ -23.1412 \end{vmatrix}, \text{ and } |X| = \begin{vmatrix} x \\ y \end{vmatrix}$$

   Solve as $|X| = |A|^{-1} \cdot |B| = \begin{vmatrix} 26.103 \\ 67.449 \end{vmatrix}$

b. Maximum production of 500 eggs occurs when the relative humidity is 67.449 % and the constant temperature is $26.103°C$.

19. a. $L(w, s) = 10.65 + 1.13w + 1.04s - 5.83ws$
Setting both partial derivatives of L equal to 0 gives the following system of equations:
$$\begin{cases} L_w: & 1.13 - 5.83s = 0 \quad [1] \\ L_s: & 1.04 - 5.83w = 0 \quad [2] \end{cases}$$
Solving the system of partial derivatives yields the critical point $\boxed{(0.178, 0.194, 10.852\%)}$.
- Algebra help for solving the system of partial derivatives:
  | Solving equation [1] yields $s \approx 0.194$
  | Solving equation [2] yields $w \approx 0.178$
  | $L(0.178, 0.194) \approx 10.852$

- Matrices are not used to solve this system because the partial derivatives are trivial.

b. Using the Determinant Test to identify the type of critical point:
The second partial derivatives are $L_{ww} = 0$, $L_{ss} = 0$, and $L_{ws} = L_{sw} = 5.83$
$$D = \begin{vmatrix} 0 & -0.583 \\ -0.583 & 0 \end{vmatrix} \approx -0.340$$
Because $D < 0$, the critical point is a $\boxed{\text{saddle point}}$.

21. a. $E(e, n) = -10.124e^3 + 21.347e^2 - 13.972e - 2.5n^2 + 2.497n + 802.2$
Setting both partial derivatives of E equal to 0 gives the following system of equations:
$$\begin{cases} E_e: & -30.372e^2 + 42.694e - 13.972 = 0 \quad [1] \\ E_n: & -5n + 2.497 = 0 \quad [2] \end{cases}$$
Solving the system of partial derivatives yields the two critical points
$\boxed{(0.519 \text{ miles}, 0.499 \text{ miles}, 799.907 \text{ feet}) \text{ and } (0.887 \text{ miles}, 0.499 \text{ miles}, 800.160 \text{ feet})}$.
- Algebra help for solving the system of partial derivatives:
  | Solving equation [1] yields the two solutions
  | $\quad e \approx 0.519$
  | $\quad e \approx 0.887$
  | Solving equation [2] yields $n = 0.499$
  | Because solutions for e and n are found independent of each other, the solution for n is paired with each of the solutions for e giving two distinct points.
  | $E(0.519, 0.499) \approx 799.907$
  | $E(0.887, 0.499) \approx 800.160$

- Matrices are not used to solve this system because the partial derivatives are not both linear.

b.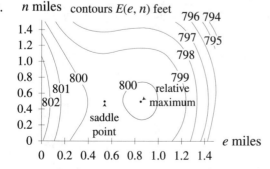

(0.519 miles, 0.499 miles, 799.907 feet) is a saddle point.
(0.887 miles, 0.499 miles, 800.160 feet) is a relative maximum point.

23. Answers will vary but might be similar to the following:
In order to determine algebraically whether a function, f, with two input variables, x and y, has a critical point:
Set both partial derivatives, f_x and f_y, equal to zero.

Calculate the solution(s) to the system of partial derivative equations and evaluate the multivariable function $f(x, y)$ at each solution. Any points located in this manner are critical points.
To determine the type of critical point found use the Determinant Test as follows.
For each critical point...
Evaluate the second partial derivatives, f_{xx}, f_{yy}, f_{xy}, and f_{yx}, and calculate the determinant of the second partials matrix, $D = f_{xx} \cdot f_{yy} - f_{xy} \cdot f_{yx}$.

If D is negative, then the critical point is a saddle point.
If D is positive, then the critical point is either a relative maximum or a relative minimum.
If $f_{xx} > 0$, the critical point is a relative minimum; if $f_{xx} < 0$, the critical point is a relative maximum.
If $D = 0$, the determinant test does not give any information about the critical point.

Section 8.3 Optimization Under Constraint (pages 607–610)

1. a. No relative extreme points are suggested by the contour graph because there are no closed contour curves.

   b.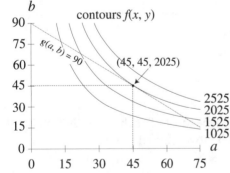

   The graph of the constraint equation $g = 90$ appears to touch but not cross the 2025-contour of f. Because 2025 is larger than any other contour level of f that the line $g = 90$ crosses, the point where the graph of g and the 2025-contour meet is the constrained maximum point: (45, 45, 2025).

3. a.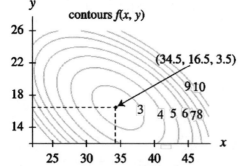

The center of the concentric contour curves appears to be (34.5, 16.5). The level of f at that point is less than 3 but greater than 2. The relative minimum point appears to be (34.5, 16.5, 3.5).

b.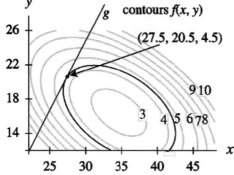

The graph of the constraint function g would touch but not cross a contour curve drawn between the 4-contour and the 5-contour of f (near the 4.25 level). Because 4.25 is smaller than any other contour level of f that the line g crosses, the point where the graph of g and the 4.25-contour meet is the constrained minimum point: (27.5, 20.5, 4.25).

5. a. The constrained system of equations:
$$\begin{cases} \text{optimize } f(r, p) = 2r^2 + rp - p^2 + p \\ \text{subject to } g(r, p) = 2r + 3p = 1 \end{cases}$$

The partial derivatives of f and g:
$f_r = 4r + p$
$f_p = r - 2p + 1$
$g_r = 2$
$g_p = 3$

The Lagrange system:
$$\begin{cases} 4r + p = 2\lambda \\ r - 2p + 1 = 3\lambda \\ 2r + 3p = 1 \end{cases}$$

b. Solving the Lagrange system yields the point $\boxed{(-0.0625, 0.375, 0.21875)}$.

- Algebra help for solving the Lagrange system:
$$\begin{cases} 4r + p = 2\lambda & [1] \\ r - 2p + 1 = 3\lambda & [2] \\ 2r + 3p = 1 & [3] \end{cases}$$

Solving both equations [1] and [2] for λ:
$$\begin{cases} \lambda = \dfrac{4r + p}{2} & [1a] \\ \lambda = \dfrac{r - 2p + 1}{3} & [2a] \end{cases}$$

Setting RHS of [1a] equal to RHS of [2a] and simplifying:
$$\dfrac{4r + p}{2} = \dfrac{r - 2p + 1}{3}$$
$$3(4r + p) = 2(r - 2p + 1)$$
$$12r + 3p = 2r - 4p + 2$$
$$10r + 7p = 2 \quad [4]$$

The Lagrange system simplifies to
$$\begin{cases} 10r + 7p = 2 & [4] \\ 2r + 3p = 1 & [3] \end{cases}$$

Solving equation [3] for p:
$$2r + 3p = 1$$
$$2r = 1 - 3p$$
$$r = \dfrac{1 - 3p}{2} \quad [5]$$

Substituting equation [5] into equation [4] gives
$$10\left(\dfrac{1 - 3p}{2}\right) + 7p = 2 \quad [6]$$

Solving equation [6] yields $p = 0.375$
Substituting $p = 0.375$ into equation [5] gives $r = -0.0625$
Substituting $p = 0.375$ and $r = -0.0625$ into equation [1a] gives $\lambda = 0.0625$.
$f(-0.0625, 0.375) = 0.21875$

- Alternate solution using matrices:

Rewrite the Lagrange system as
$$\begin{cases} 4r + p - 2\lambda = 0 \\ r - 2p - 3\lambda = -1 \\ 2r + 3p \quad\quad = 1 \end{cases}$$

Represent the system using matrices:
$$|A| = \begin{vmatrix} 4 & 1 & -2 \\ 1 & -2 & -3 \\ 2 & 3 & 0 \end{vmatrix}, \ |B| = \begin{vmatrix} 0 \\ -1 \\ 1 \end{vmatrix}, \ \text{and } |X| = \begin{vmatrix} r \\ p \\ \lambda \end{vmatrix}$$

Solve as $|X| = |A|^{-1} \cdot |B| = \begin{vmatrix} -0.0625 \\ 0.375 \\ 0.0625 \end{vmatrix}$

c. Identifying the type of constrained critical point by comparing output values of f at close points on the constraint $g(r, p) = 1$:

The table below is constructed using nearby values for p and calculating corresponding values for r using $r = \dfrac{1-3p}{2}$ (that is, $g = 1$ solved for r).

	r	p	$f(r, p)$	classification
checkpoint ($p_0 - 0.005$)	−0.055	0.370	0.2188	
critical point (p_0)	**−0.0625**	**0.375**	**0.21875**	constrained minimum point
checkpoint ($p_0 + 0.005$)	−0.07	0.380	0.2188	

7. a. The constrained system of equations:
$$\begin{cases} \text{optimize } f(x, y) = 80x + 5y^2 \\ \text{subject to } g(x, y) = 2x + 2y = 1.4 \end{cases}$$

The partial derivatives of f and g:
$f_x = 80$
$f_y = 10y$
$g_x = 2$
$g_y = 2$

The Lagrange system:
$$\begin{cases} 80 = 2\lambda \\ 10y = 2\lambda \\ 2x + 2y = 1.4 \end{cases}$$

b. Solving the Lagrange system yields the point $(-7.3, 8, -264)$.

• Algebra help for solving the Lagrange system:
$$\begin{cases} 80 = 2\lambda & [1] \\ 10y = 2\lambda & [2] \\ 2x + 2y = 1.4 & [3] \end{cases}$$

Setting LHS of [1] equal to LHS of [2] and simplifying:
$10y = 80$
$y = 8$

Substituting $y = 8$ into equation [3] gives
$2x + 2(8) = 1.4$ [4]

Solving equation [4] yields $x = -7.3$.
Solving equation [1] yields $\lambda = 40$.
$f(-7.3, 8) = -264$

- Alternate solution using matrices:

   Rewrite the Lagrange system as
   $$\begin{cases} 2\lambda = 80 \\ 10y - 2\lambda = 0 \\ 2x + 2y = 1.4 \end{cases}$$

   Represent the system using matrices:
   $$|A| = \begin{vmatrix} 0 & 0 & 2 \\ 0 & 10 & -2 \\ 2 & 2 & 0 \end{vmatrix}, |B| = \begin{vmatrix} 80 \\ 0 \\ 1.4 \end{vmatrix}, \text{ and } |X| = \begin{vmatrix} x \\ y \\ \lambda \end{vmatrix}$$

   Solve as $|X| = |A|^{-1} \cdot |B| = \begin{vmatrix} -7.3 \\ 8 \\ 40 \end{vmatrix}$

c. Identifying the type of constrained critical point by comparing output values of f at close points on the constraint $g(x, y) = 1.4$:

   The table below is constructed using nearby values for y and calculating corresponding values for x using $x = \dfrac{1.4 - 2y}{2}$ (that is, $g = 1.4$ solved for x).

	x	y	$f(x, y)$	classification
checkpoint $(y_0 - 0.1)$	−7.2	7.9	−263.95	
critical point (y_0)	−7.3	8	−264	constrained maximum point
checkpoint $(y_0 + 0.1)$	−7.4	8.1	−263.95	

9. a. The constrained system of equations:
   $$\begin{cases} \text{optimize } f(x, y) = x^2 y \\ \text{subject to } g(x, y) = x + y = 16 \end{cases}$$

   The partial derivatives of f and g:
   $f_x = 2xy$
   $f_y = x^2$
   $g_x = 1$
   $g_y = 1$

   The Lagrange system:
   $$\begin{cases} 2xy = \lambda \\ x^2 = \lambda \\ x + y = 16 \end{cases}$$

   b. Solving the Lagrange system yields two points:
   (0, 16, 0)
   (10.667, 5.333, 606.815)

- Algebra help for solving the Lagrange system:

$$\begin{cases} 2xy = \lambda & [1] \\ x^2 = \lambda & [2] \\ x + y = 16 & [3] \end{cases}$$

Setting LHS of equation [1] equal to LHS of equation [2] gives

$$2xy = x^2$$
$$2xy - x^2 = 0$$
$$x(2y - x) = 0 \quad \text{by factoring}$$

Solving $x(2y - x) = 0$ for x leads to two cases:

$x = 0$ [case 1]
$x = 2y$ [case 2]

Case 1:
Substituting $x = 0$ into equation [3] gives $y = 16$ and equation [2] gives $\lambda = 0$.
$f(0, 16) = 0$

Case 2:
Substituting $x = 2y$ into equation [3] gives

$$2y + y = 16$$
$$3y = 16$$
$$y = \frac{16}{3} \quad \rightarrow \quad y \approx 5.333$$

Substituting $y \approx 5.333$ into $x = 2y$ gives $x \approx 10.667$.
Substituting $x \approx 10.667$ into equation [2] gives $\lambda \approx 113.778$
$f(10.667, 5.333) \approx 606.815$

- Matrices are not used to solve this system because not all of the equations are linear.

c. Identifying the type of constrained critical point by comparing output values of f at close points on the constraint $g(x, y) = 16$:

The table below is constructed using nearby values for y and calculating corresponding values for x using $x = 16 - y$ (that is, $g = 16$ solved for x).

	x	y	$f(x, y)$	classification
checkpoint ($y_1 - 0.1$)	0.100	15.9	0.159	
critical point (y_1)	0	16	0	constrained minimum point
checkpoint ($y_1 + 0.1$)	−0.100	16.1	0.161	
checkpoint ($y_2 - 0.1$)	10.767	5.233	606.654	
critical point (y_2)	10.667	5.333	606.815	constrained maximum point
checkpoint ($y_2 + 0.1$)	10.567	5.433	606.656	

11. a. The constrained system of equations:
$$\begin{cases} \text{optimize } f(x, y) = 100x^{0.8}y^{0.2} \\ \text{subject to } g(x, y) = 2x + 4y = 100 \end{cases}$$
The partial derivatives of f and g:
$$f_x = 80x^{-0.2}y^{0.2}$$
$$f_y = 20x^{0.8}y^{-0.8}$$
$$g_x = 2$$
$$g_y = 4$$
The Lagrange system:
$$\begin{cases} 80x^{-0.2}y^{0.2} = 2\lambda \\ 20x^{0.8}y^{-0.8} = 4\lambda \\ 2x + 4y = 100 \end{cases}$$

b. Solving the Lagrange system yields the point $(40, 5, 2639.016)$.

• Algebra help for solving the Lagrange system:
$$\begin{cases} 80x^{-0.2}y^{0.2} = 2\lambda & [1] \\ 20x^{0.8}y^{-0.8} = 4\lambda & [2] \\ 2x + 4y = 100 & [3] \end{cases}$$
Reducing equations [1] and [2] gives
$$40x^{-0.2}y^{0.2} = \lambda \quad [1a]$$
$$5x^{0.8}y^{-0.8} = \lambda \quad [2a]$$
Setting LHS of equation [1a] equal to LHS of equation [2a] gives
$$40x^{-0.2}y^{0.2} = 5x^{0.8}y^{-0.8} \quad [4]$$
Solving equation [4] for x:
$$40x^{-0.2}y^{0.2} = 5x^{0.8}y^{-0.8}$$
$$8y^{0.2}y^{0.8} = x^{0.8}x^{0.2}$$
$$8y = x \quad [5]$$
Substituting equation [5] into equation [3] gives
$$2(8y) + 4y = 100 \quad [6]$$
Solving equation [6] yields $y = 5$.
Substituting $y = 5$ into equation [5] gives $x = 40$.
Substituting $x = 40$ and $y = 5$ into equation [1a] gives $\lambda \approx 26.390$
$f(40, 5) \approx 2639.016$

• Matrices are not used to solve this system because not all of the equations are linear.

c. Identifying the type of constrained critical point by comparing output values of f at close points on the constraint $g(x, y) = 100$:
The table below is constructed using nearby values for x and calculating corresponding values for y using $y = \dfrac{100 - 2x}{4}$ (that is, $g = 100$ solved for y).

	x	y	$f(x, y)$	classification
checkpoint $(x_0 - 1)$	39	5.5	2635.871	
critical point (x_0)	**40**	**5**	**2639.016**	constrained maximum point
checkpoint $(x_0 + 1)$	41	4.4	2635.540	

13. a. The constrained system of equations:
$$\begin{cases} \text{optimize } A(r, n) = 0.1r^2 n \\ \text{subject to } g(r, n) = 12r + 6n = 504 \end{cases}$$

The partial derivatives of A and g:

$A_r = 0.2rn$

$A_n = 0.1r^2$

$g_r = 12$

$g_n = 6$

The Lagrange system:
$$\begin{cases} 0.2rn = 12\lambda \\ 0.1r^2 = 6\lambda \\ 12r + 6n = 504 \end{cases}$$

Solving the Lagrange system yields two points:

(0, 84, 0) is not useful in context.

(28, 28, 2195.2) corresponds to a constrained maximum point.

(For algebra help see the bulleted sections following the main solution.)

Optimal allocation: radio advertising: (28 ads)·($12 per ad) = $\boxed{\$336}$

newspaper advertising: (28 ads)·($6 per ad) = $\boxed{\$168}$

• Algebra help for solving the Lagrange system:
$$\begin{cases} 0.2rn = 12\lambda & [1] \\ 0.1r^2 = 6\lambda & [2] \\ 12r + 6n = 504 & [3] \end{cases}$$

Dividing equation [1] by 2 gives

$0.1rn = 6\lambda$ [1a]

Setting LHS of equation [1a] equal to LHS of equation [2] gives

$0.1rn = 0.1r^2$ [4]

Setting LHS of equation [1a] equal to LHS of equation [2] gives

$0.1rn = 0.1r^2$

$rn = r^2$

$rn - r^2 = 0$

$r(n - r) = 0$ by factoring

Solving $r(n - r) = 0$ for r leads to two cases:

$r = 0$ [case 1]

$r = n$ [case 2]

| | Case 1:
| | Substituting $r = 0$ into equation [3] gives $r = 84$ and equation [2] gives $\lambda = 117.6$.
| | $A(0, 84) = 0$
| | Case 2:
| | Substituting $r = n$ into equation [3] gives
| | $\quad 12n + 6n = 504$
| | $\quad\quad\quad 18n = 504$
| | $\quad\quad\quad\quad n = 28$
| | Substituting $n = 28$ into $r = n$ gives $r = 28$.
| | Substituting $r = 28$ into equation [2] gives $\lambda \approx 13.067$
| | $A(28, 28) = 2195.2$

•Matrices are not used to solve this system because not all of the equations are linear. Identifying the type of constrained critical point by comparing output values of A at close points on the constraint $g(n, r) = 504$:

The table below is constructed using nearby values for n and calculating corresponding values for r using $r = \dfrac{504 - 6n}{12}$ (that is, $g = 504$ solved for r).

	n	r	$A(n, r)$	classification
checkpoint $(n_1 - 0.1)$	0.1	42.05	−17.682	
critical point (n_1)	**0**	**42**	**0**	constrained saddle point
checkpoint $(n_1 + 0.1)$	−0.1	41.95	17.598	
checkpoint $(n_2 - 0.1)$	27.9	28.05	2195.179	
critical point (n_2)	**28**	**28**	**2195.2**	constrained maximum point
checkpoint $(n_2 + 0.1)$	28.1	27.95	2195.179	

b. $A(28, 28) \approx \boxed{2195 \text{ responses}}$

c. $\Delta A \cdot \lambda \approx 26 \cdot 13.067$

$\quad\quad \approx 339.733 \quad \rightarrow \quad \boxed{340 \text{ responses}}$

15. a. $\dfrac{2}{3}(320) + x + y + 63 = 320$

b. The constrained system of equations:
$$\begin{cases} \text{optimize } q(x, y) = 50x^{0.7}y^{0.3} \\ \text{subject to } g(x, y) = \dfrac{2}{3}(320) + x + y + 63 = 320 \end{cases}$$

The partial derivatives of q and g:
$q_x = 35x^{-0.3}y^{0.3}$
$q_y = 15x^{0.7}y^{-0.7}$
$g_x = 1$
$g_y = 1$

8.3 Optimization under Constraints
Solutions to Odd-Numbered Problems

The Lagrange system:
$$\begin{cases} 35x^{-0.3}y^{0.3} = \lambda \\ 15x^{0.7}y^{-0.7} = \lambda \\ \dfrac{2}{3}(320) + x + y + 63 = 320 \end{cases}$$

Set the first two equations equal to each other: $35x^{-0.3}y^{0.3} = 15x^{0.7}y^{-0.7}$

Solve for y:

$$\dfrac{35x^{-0.3}y^{0.3}}{35x^{-0.3}y^{-0.7}} = \dfrac{15x^{0.7}y^{-0.7}}{35x^{-0.3}y^{-0.7}}$$

$$y = \dfrac{15x}{35}$$

Substitute for y in the constraint equation: $\dfrac{2}{3}(320) + x + \dfrac{15x}{35} + 63 = 320$

Solve for x: $x = 30.557$

Evaluate y: $y = \dfrac{15(30.557)}{35} = 13.1$ $\rightarrow$ $\boxed{\$13{,}100}$

Verifying the type of constrained critical point by comparing output values of q at close points on the constraint $g(x, y) = 200$:

The table below is constructed using nearby values for x and calculating corresponding values for y using $y = \dfrac{200 - 10x}{3}$ (that is, $g = 200$ solved for y).

	x	y	$q(x, y)$	classification
checkpoint $(x_0 - 0.557)$	30	13.667	1184.823	
critical point (x_0)	**30.557**	**13.1**	**1185.029**	constrained maximum point
checkpoint $(x_0 + 0.543)$	31	12.667	1185.010	

17. a. $f(x, y) = -125.48 + 4.26x + 4.85y - 0.05x - 0.14y^2$

The partial derivatives:
$f_x = 4.26 - 0.1x$
$f_y = 4.85 - 0.28y$

The system of partial derivatives:
$$\begin{cases} 4.26 - 0.1x = 0 \\ 4.85 - 0.28y = 0 \end{cases}$$

Solving the system yields $x = 42.6\%$ and $y = 17.3\%$

$f(42.6,\ 17.3) = \boxed{7.3}$

Using the Determinant Test to classify the critical point:

The second derivatives are $f_{xx} = -0.1$, $f_{yy} = -0.28$, and $f_{xy} = f_{yx} = 0$

$$D = \begin{vmatrix} -0.1 & 0 \\ 0 & -0.28 \end{vmatrix} = 0.028$$

Because $D > 0$ and $f_{xx} < 0$, the critical point is a relative maximum point.

b. The constrained system of equations:
$$\begin{cases} \text{optimize } f(x, y) = -125.48 + 4.26x + 4.85y - 0.05x^2 - 0.14y^2 \\ \text{subject to } g(x, y) = x + y = 55 \end{cases}$$

The partial derivatives of A and g:
$f_x = 4.26 - 0.1x$
$f_y = 4.85 - 0.28y$
$g_x = 1$
$g_y = 1$

The Lagrange system:
$$\begin{cases} 4.26 - 0.1x = 1\lambda \\ 4.85 - 0.28y = 1\lambda \\ x + y = 55 \end{cases}$$

Set the first and second equations equal to each other: $4.26 - 0.1x = 4.85 - 0.28y$

Solve the system: $\begin{cases} 4.26 - 0.1x = 4.85 - 0.28y \\ x + y = 55 \end{cases}$

$x = 39\%, y = 16\%$, and $\lambda = 0.38$

$f(39, 16) = \boxed{6.365}$

c. Verifying the type of constrained critical point by comparing output values of f at close points on the constraint $g(x, y) = 55$:

The table below is constructed using nearby values for x and calculating corresponding values for y using $y = 55 - x$ (that is, $g = 55$ solved for y).

	x	y	$f(x, y)$	classification
checkpoint $(x_0 - 1)$	38	17	6.190	
critical point (x_0)	39	16	6.365	constrained maximum point
checkpoint $(x_0 + 1)$	40	15	6.170	

19. a. Worker expenditure = $\dfrac{\dfrac{\$7.50}{\text{hour}} \cdot 100L \text{ hours}}{1000} = \$0.75$ thousand

The constraint is $g(L, K) = 0.75L + K$ thousand dollars where L hundred hours is the size of the labor force and K thousand dollars is the capital investment.

The constrained system of equations:
$$\begin{cases} \text{optimize } f(L, K) = 10.5463 L^{0.3} K^{0.5} \\ \text{subject to } g(L, K) = 0.75L + K = 15 \end{cases}$$

The partial derivatives of A and g:
$f_L = 3.16389 L^{-0.7} K^{0.5}$
$f_K = 5.27315 L^{0.3} K^{-0.5}$
$g_L = 0.75$
$g_K = 1$

The Lagrange system:

8.3 Optimization under Constraints
Solutions to Odd-Numbered Problems

$$\begin{cases} 3.16389L^{-0.7}K^{0.5} = 0.75\lambda \\ 5.27315L^{0.3}K^{-0.5} = 1\lambda \\ 0.75L + K = 15 \end{cases}$$

Solve the first and second equations for λ and set equal: $\dfrac{3.16389L^{-0.7}K^{0.5}}{0.75} = \dfrac{5.27315L^{0.3}K^{-0.5}}{1}$

Solving for K yields $K = 1.25L$
Substituting in the constraint and solving for yields
$\boxed{L = 7.5 \text{ hundred hours and } K = \$9.375 \text{ thousand}}$
$\lambda = 3.152$ radios per thousand dollars
$f(7.5, 9.375) = \boxed{59.102 \text{ radios}}$ radios

b. Verifying the type of constrained critical point by comparing output values of f at close points on the constraint $g(L, K) = 15$:

The table below is constructed using nearby values for L and calculating corresponding values for K using $K = 15 - 0.75L$ (that is, $g = 15$ solved for K).

	L	K	f(L, K)	classification
checkpoint $(L_0 - 0.5)$	7	9.750	59.038	
critical point (L_0)	**7.5**	**9.375**	**59.102**	constrained maximum point
checkpoint $(L_0 + 0.5)$	8	9.000	59.040	

c. The marginal productivity of money is 3.152 radios per thousand dollars. An increase in the budget of $\$1000$ will result in an increase in output of about 3 radios.

21. a. $R(s, p) = (50 + s)p$ dollars where s is the number of students in excess of 50 and $\$p$ is the price per student
b. $g(s, p) = p + 10s = 1200$
c. The constrained system of equations:
$$\begin{cases} \text{optimize: } R(s, p) = 50p + sp \\ \text{subject to: } g(s, p) = p + 10s = 1200 \end{cases}$$
The partial derivatives of R and g:
$R_s = p$
$R_p = 50 + s$
$g_s = 10$
$g_p = 1$
The Lagrange system:
$$\begin{cases} p = 10\lambda \\ 50 + s = 1\lambda \\ p + 10s = 1200 \end{cases}$$
Solve the first and second equations for λ and set equal:
$\dfrac{p}{10} = 50 + s$

Substitute $p = 500 + 10s$ in the third equation and solve:
$s = 35$ people, $p = \$850$, and $R(35, 850) = \$72,250$

The travel agency's revenue will be maximized when 85 people go on the cruise and pay $850 each. The revenue will be $72.250.

Verifying the type of constrained critical point by comparing output values of R at close points on the constraint $g(s, p) = 1200$:

The table below is constructed using nearby values for s and calculating corresponding values for p using $p = 1200 - 10s$ (that is, $g = 1200$ solved for p).

	s	p	$R(s, p)$	classification
checkpoint $(s_0 - 0.5)$	34.5	855	72,247.5	
critical point (s_0)	**35**	**850**	**72,250**	constrained maximum point
checkpoint $(s_0 + 0.5)$	35.5	845	72,247.5	

· Using matrices to solve the Lagrange system:

$$[A] = \begin{bmatrix} 1 & 0 & -10 \\ 0 & 1 & -1 \\ 1 & 10 & 0 \end{bmatrix}, [X] = \begin{bmatrix} p \\ s \\ \lambda \end{bmatrix}, [B] = \begin{bmatrix} 0 \\ -50 \\ 1200 \end{bmatrix}$$

Solving, $[X] = [A]^{-1} \cdot [B] = \begin{bmatrix} 850 \\ 35 \\ 85 \end{bmatrix}$

23. a. $A(l, w) = lw$
    b. $2l + 2w = 206$
    c. The constrained system of equations:
    $$\begin{cases} \text{optimize } A(l, w) = lw \\ \text{subject to } g(l, w) = 2l + 2w = 206 \end{cases}$$
    The partial derivatives of A and g:
    $A_l = w$
    $A_w = l$
    $g_l = 2$
    $g_w = 2$
    The Lagrange system:
    $$\begin{cases} w = 2\lambda \\ l = 2\lambda \\ 2l + 2w = 206 \end{cases}$$
    Setting the first two equations equal yields $l = w$.
    Substituting $l = w$ into the third equation and solving yields
    $l = 51.5$
    $w = 51.5$
    $A(51.5, 51.5) = 2,652.25$

The critical point is (51.5 feet, 51.5 feet, 2652.25 square feet)

The corral with a perimeter of 206 feet and greatest area is 51.5 feet square with an area of 2652.25 square feet.

Verifying the type of constrained critical point by comparing output values of A at close points on the constraint $g(w, l) = 206$:

The table below is constructed using nearby values for l and calculating corresponding values for w using $w = \dfrac{206 - 2l}{2}$ (that is, $g = 206$ solved for w).

	l	w	$A(l, w)$	classification
checkpoint $(l_0 - 0.5)$	51	52	2652.00	
critical point (l_0)	**51.5**	**51.5**	**2652.25**	constrained maximum point
checkpoint $(l_0 + 0.5)$	52	51	2652.00	

25. Answers will vary but might be similar to the following:

The condition $\dfrac{f_x}{f_y} = \dfrac{g_x}{g_y}$ is equivalent to guaranteeing that the slope of the extreme-contour curve is the same as the slope of the constraint curve at their point of intersection.

Section 8.4 Least-Squares Optimization (pages 615–617)

1. a. $f(a, b) = SSE = [7 - (a \cdot 1 + b)]^2 + [11 - (a \cdot 6 + b)]^2 + [19 - (a \cdot 12 + b)]^2$

   $\boxed{f(a, b) = (7 - a - b)^2 + (11 - 6a - b)^2 + (19 - 12a - b)^2}$

   Alternate form: $f(a, b) = 531 + 181a^2 - 602a + 38ab - 74b + 3b^2$

   b. first partials (by the Chain Rule):

   $\dfrac{\partial f}{\partial a} = 2(7 - a - b) \cdot (-1) + 2(11 - 6a - b) \cdot (-6) + 2(19 - 12a - b) \cdot (-12)$

   $\boxed{\dfrac{\partial f}{\partial a} = 362a - 602 + 38b}$

   $\dfrac{\partial f}{\partial b} = 2(7 - a - b) \cdot (-1) + 2(11 - 6a - b) \cdot (-1) + 2(19 - 12a - b) \cdot (-1)$

   $\boxed{\dfrac{\partial f}{\partial b} = 38a - 74 + 6b}$

   determinant of second partials (each second partial is constant):

   $\begin{vmatrix} 362 & 38 \\ 38 & 6 \end{vmatrix} = (362)(6) - (38)(38) = 728$

   (The positive determinant and positive f_{aa} guarantee that the point found is a minimum.)

c. Solving the system of partials $\begin{cases} 362a - 602 + 38b = 0 & [1] \\ 38a - 74 + 6b = 0 & [2] \end{cases}$

Solving equation [2] for b yields $b = \dfrac{-38a + 74}{6}$

algebra help: $38a - 74 + 6b = 0$
$$6b = -38a + 74$$
$$b = \dfrac{-38a + 74}{6}$$

Substituting $b = \dfrac{-38a + 74}{6}$ into equation [1] gives

$$362a - 602 + 38 \cdot \dfrac{-38a + 74}{6} = 0 \quad [3]$$

Solving equation [3] for a yields $a \approx 1.099$

Substituting $a \approx 1.099$ into $b = \dfrac{-38a + 74}{6}$ gives $b \approx 5.374$

Solution: $\begin{cases} a \approx 1.099 \\ b \approx 5.374 \end{cases}$

d. $y \approx 1.099x + 5.374$

3. a. $f(a, b) = SSE = [3 - (a \cdot 0 + b)]^2 + [2 - (a \cdot 10 + b)]^2 + [1 - (a \cdot 20 + b)]^2$

$f(a, b) = (3 - b)^2 + (2 - 10a - b)^2 + (1 - 20a - b)^2$

Alternate form: $f(a, b) = 14 + 500a^2 - 80a + 60ab - 12b + 3b^2$

b. first partials (by the Chain Rule):

$\dfrac{\partial f}{\partial a} = 2(3 - b) \cdot (0) + 2(2 - 10a - b) \cdot (-10) + 2(1 - 20a - b) \cdot (-20)$
$\phantom{\dfrac{\partial f}{\partial a}} = 1000a - 80 + 60b$

$\dfrac{\partial f}{\partial b} = 2(3 - b) \cdot (-1) + 2(2 - 10a - b) \cdot (-1) + 2(1 - 20a - b) \cdot (-1)$
$\phantom{\dfrac{\partial f}{\partial b}} = 60a - 12 + 6b$

Solving the system of partials $\begin{cases} 1000a - 80 + 60b = 0 & [1] \\ 60a - 12 + 6b = 0 & [2] \end{cases}$:

Solving equation [2] for b yields $b = 2 - 10a$

algebra help: $60a - 12 + 6b = 0$
$$6b = -60a + 12$$
$$b = \dfrac{-60a + 12}{6}$$
$$b = -10a + 2$$

Substituting $b = 2 - 10a$ into equation [1] gives

$1000a - 80 + 60(2 - 10a) = 0 \quad [3]$

Solving equation [3] for a yields $a = -0.1$

Substituting $a = -0.1$ into $b = 2 - 10a$ gives $b = 3$

8.4 Least-Squares Optimization
Solutions to Odd-Numbered Problems

Solution: $\begin{cases} a = -0.1 \\ b = 3 \end{cases}$

Determinant of second partials matrix (each second partial is constant):

$$\begin{vmatrix} 1000 & 60 \\ 60 & 6 \end{vmatrix} = (1000)(6) - (60)(60) = 2400$$

(The positive determinant and positive f_{aa} guarantee that the point found is a minimum.)

Minimum SSE: $f(-0.1, 3) = \boxed{0}$

SSE is zero indicating that the line with parameters $a = -0.1$ and $b = 3$ passes through each data point.

c. $h(x) = -0.1x + 3$ percent gives the percentage of homes with incomplete kitchens in the western United States x years after 1970, data from $0 \leq x \leq 20$.

d. Solving $h(x) = 0$ yields $x = 30 \quad \rightarrow \quad \boxed{2000}$

5. a. y dollars

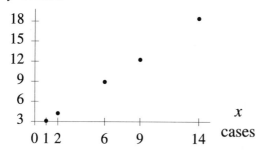

b. $C(x) \approx 1.176x + 1.880$ dollars gives the production cost for x cases of 7 mm aluminum ball bearings, $1 \leq x \leq 14$.

y dollars

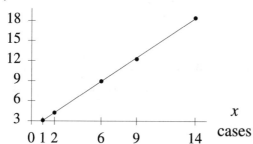

c.

data x	data y	model $C(x)$	deviation $y - C(x)$	squared error
1	3.10	3.056	0.044	0.001925
2	4.25	4.232	0.018	0.000310
6	8.95	8.937	0.013	0.000156
9	12.29	12.466	−0.176	0.031084
14	18.45	18.348	0.102	0.010472
			SSE ≈	$\boxed{0.044}$

7. $f(a, b) = SSE = [5.5-(a \cdot 0+b)]^2 + [5-(a \cdot 5+b)]^2 + [4.8-(a \cdot 8+b)]^2 + [4.6-(a \cdot 10+b)]^2$
$= (5.5-b)^2 + (5-5a-b)^2 + (4.8-8a-b)^2 + (4.6-10a-b)^2$

Alternate form: $f(a, b) = 99.45 + 189a^2 - 218.8a + 46ab - 39.8b + 4b^2$

First partial derivatives (by the Chain Rule):

$\dfrac{\partial f}{\partial a} = 0 + 2(5-5a-b)(-5) + 2(4.8-8a-b)(-8) + 2(4.6-10a-b)(-10)$
$= 378a - 218.8 + 46b$

$\dfrac{\partial f}{\partial b} = 2(5.5-b)(-1) + 2(5-5a-b)(-1) + 2(4.8-8a-b)(-1) + 2(4.6-10a-b)(-1)$
$= 46a - 39.8 + 8b$

Solving the system of partials $\begin{cases} 378a - 218.8 + 46b = 0 & [1] \\ 46a - 39.8 + 8b = 0 & [2] \end{cases}$.

Solving equation [2] for b yields $b = -5.75a + 4.975$
algebra help: $46a - 39.8 + 8b = 0$
$8b = -46a + 39.8$
$b = \dfrac{-46a + 39.8}{8}$
$b = -5.75a + 4.975$

Substituting $b = -5.75a + 4.975$ into equation [1] gives
$378a - 218.8 + 46(-5.75a + 4.975) = 0$ [3]

Solving equation [3] for a yields $a \approx -0.089$
Substituting $a \approx -0.089$ into $b = -5.75a + 4.975$ gives $b \approx 5.484$

Solution: $\begin{cases} a \approx -0.089 \\ b \approx 5.484 \end{cases}$

Determinant of second partials matrix (each second partial is constant):

$\begin{vmatrix} 378 & 46 \\ 46 & 8 \end{vmatrix} = (378)(8) - (46)(46) = 908$

(The positive determinant and positive f_{aa} guarantee that the point found is a minimum.)

$\boxed{E(x) \approx -0.089x + 5.484 \text{ million experiments gives the number of animal experiments in England } x \text{ years since 1970, data from } 0 \leq x \leq 10.}$

9. a. billion people

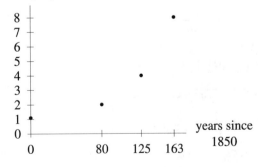

A scatter plot suggests a concave up, increasing function—possibly exponential.

b.

year since 1850	natural log of population
0	0.095
80	0.693
125	1.386
163	2.079

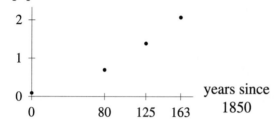

c. $f(a, b) = SSE = [\ln 1.1 - (a \cdot 0 + b)]^2 + [\ln 2.0 - (a \cdot 80 + b)]^2 + [\ln 4.0 - (a \cdot 125 + b)]^2$
$$+ [\ln 8.0 - (a \cdot 163 + b)]^2$$
$$= (\ln 1.1 - b)^2 + (\ln 2.0 - 80a - b)^2 + (\ln 4.0 - 125a - b)^2 + (\ln 8.0 - 163a - b)^2$$

Alternate form: $f(a, b) \approx 6.735 + 48594a^2 - 1135.375a + 736ab - 8.508b + 4b^2$

first partials (by the Chain Rule):

$$\frac{\partial f}{\partial a} = 0 + 2(\ln 2.0 - 80a - b)(-80) + 2(\ln 4.0 - 125a - b)(-125) + 2(\ln 8.0 - 163a - b)(-163)$$
$$\approx 97{,}188a - 1135.375 + 736b$$

$$\frac{\partial f}{\partial b} = 2(\ln 1.1 - b)(-1) + 2(\ln 2.0 - 80a - b)(-1) + 2(\ln 4.0 - 125a - b)(-1)$$
$$+ 2(\ln 8.0 - 163a - b)(-1)$$
$$\approx 736a - 8.508 + 8b$$

Solving the system of partials $\begin{cases} 97{,}188a - 1135.375 + 736b = 0 & \textbf{[1]} \\ 736a - 8.508 + 8b = 0 & \textbf{[2]} \end{cases}$.

Solving equation [2] for b yields $b = 1.064 - 92a$
algebra help: $736a - 8.508 + 8b = 0$
$$8b = -736a + 8.508$$
$$b = \frac{-736a + 8.508}{8}$$
$$b = -92a + 1.064$$

Substituting $b = 1.064 - 92a$ into equation [1] gives
$$97{,}188a - 1135.375 + 736(1.064 - 92a) = 0 \quad \textbf{[3]}$$

Solving equation [3] for a yields $a \approx 0.012$
Substituting $a \approx 0.012$ into $b = 1.064 - 92a$ gives $b \approx 0.037$

Solution: $\begin{cases} a \approx 0.012 \\ b \approx 0.037 \end{cases}$

determinant of second partials (each second partial is constant):

$$\begin{vmatrix} 97{,}188 & 736 \\ 736 & 8 \end{vmatrix} = (97{,}188)(8) - (736)(736) = 235{,}808$$

(The positive determinant and positive f_{aa} guarantee that the point found is a minimum.)

$\boxed{L(x) \approx 0.012x + 0.037}$ is the linear model that fits the data for the natural log of population.

d. $y \approx e^{0.037}(e^{0.012x})$ billion people is the model for the original population data.

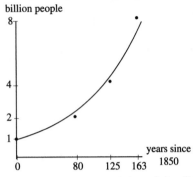

e. TI-84 generated model: $p(x) \approx 0.964(1.0120^x)$ billion people gives world population x years after 1850, data from $0 \le x \le 163$.

part d model: Rewrite $y \approx e^{0.037} e^{0.012x}$ as

$y(x) \approx 1.037(1.0117^x)$ billion people gives world population x years after 1850, data from $0 \le x \le 163$.

The parameter b is 0.043 smaller in the calculator generated model but the slightly larger percentage rate of change for that model makes the two models very close over the input interval.

Chapter 8 Review Activities

1. a.

b. Highest ozone level: 450 thousandths of a centimeter at 90°N in March.
   Lowest ozone level: 250 thousandths of a centimeter near 10°N in December.

3. a. The partial derivatives:
$$\begin{cases} T_p: & -0.50p + 0.25s + 120 = 0 \\ T_s: & 0.25p - 0.750s + 100 = 0 \end{cases}$$
Solving the system of partial derivatives yields
$$s = \boxed{256 \text{ key chains}}$$
$$p = \boxed{368 \text{ patches}}$$
(For algebra and matrix help see the bulleted sections following the main solution.)
Using the Determinant Test to verify that the critical point is a relative maximum point:
The second partial derivatives are $T_{pp} = -0.50$, $T_{ss} = 0.25$, and $T_{ps} = T_{sp} = 0.25$

$$D(368, 256) = \begin{vmatrix} -0.5 & 0.25 \\ 0.25 & -0.75 \end{vmatrix} = 0.3125$$

Because $D > 0$ and $T_{pp} < 0$, the critical point is a relative maximum point.

- Algebra help for solving the system of partial derivatives:
  Rewrite the system of partial derivatives as
  $$\begin{cases} -0.50p + 0.25s = -120 & [1] \\ 0.25p - 0.75s = -100 & [2] \end{cases}$$
  Eliminate p by multiplying equation [2] by 2 and adding the result to equation [1]:
  $$-0.50p + 0.25s = -120$$
  $$(+) \quad 0.50p - 1.50s = -200$$
  $$\overline{0 - 1.25s = -320}$$
  Solving $-1.25s = -320$ yields $s = 256$.
  Substituting $s = 256$ into equation [1] gives $-0.50p + 0.25(256) = -120$.
  Solving $-0.50p + 64 = -120$ yields $p = 368$
  $T(256, 368) = 35,880$

- Alternate solution using matrices:
  Rewrite the system of partial derivatives as
  $$\begin{cases} -0.1452w - 0.0181x = -2.4297 \\ -0.0181w + 0.0168x = 0.838 \end{cases}$$
  Represent the system using matrices:
  $$|A| = \begin{vmatrix} -0.5 & 0.25 \\ 0.25 & -0.75 \end{vmatrix}, \quad |B| = \begin{vmatrix} -120 \\ -100 \end{vmatrix}, \quad \text{and } |X| = \begin{vmatrix} p \\ s \end{vmatrix}$$
  Solve as $|X| = |A|^{-1} \cdot |B| = \begin{vmatrix} 368 \\ 256 \end{vmatrix}$

b. $T(256, 368) = \boxed{\$35,880}$

5. a. The constrained system:
$$\begin{cases} \text{optimize} & f(x, y) = 3x^2 + 2y^2 - 25.2x - 11.2y + 73.6 \\ \text{subject to} & g(x, y) = 2x + y = 6 \end{cases}$$
The four partials derivatives:
$f_x = 6x - 25.2$

$f_y = 4y - 11.2$

$g_x = 2$

$g_y = 1$

The Lagrange system:
$$\begin{cases} 6x - 25.2 = 2\lambda \\ 4y - 11.2 = 1\lambda \\ 2x + y = 6 \end{cases}$$

b. Solving the Lagrange system yields the point $(2.309, 1.382, 19.749)$.

 • Algebra help for solving the Lagrange system:
$$\begin{cases} 6x - 25.2 = 2\lambda & [1] \\ 4y - 11.2 = \lambda & [2] \\ 2x + y = 6 & [3] \end{cases}$$
Substituting equation [2] into equation [1] gives:
$6x - 25.2 = 2(4y - 11.2)$
$6x - 25.2 = 8y - 22.4$ [4]
Solving equation [3] for y gives
$y = 6 - 2x$ [5]
Substituting equation [5] into equation [4] gives
$6x - 25.2 = 8(6 - 2x) - 22.4$ [6]
Solving equation [6] yields $x \approx 2.309$.
Substituting $x \approx 2.309$ into equation [5] and solving yields $y \approx 1.382$.
Substituting $y \approx 1.382$ into equation [2] gives $\lambda = -5.673$.
$f(2.309, 1.382) = 19.749$

 • Alternate solution using matrices:
Rewrite the Lagrange system as
$$\begin{cases} 6x & -2\lambda = 25.2 \\ 4y - \lambda = 11.2 \\ 2x + y & = 6 \end{cases}$$
Represent the system using matrices:
$$|A| = \begin{vmatrix} 6 & 0 & -2 \\ 0 & 4 & -1 \\ 2 & 1 & 0 \end{vmatrix}, |B| = \begin{vmatrix} 25.2 \\ 11.2 \\ 6 \end{vmatrix}, \text{ and } |X| = \begin{vmatrix} x \\ y \\ \lambda \end{vmatrix}$$

Solve as $|X| = |A|^{-1} \cdot |B| = \begin{vmatrix} 2.309 \\ 1.382 \\ -5.673 \end{vmatrix}$

c. Identifying the type of constrained critical point by comparing output values of f at close points on the constraint $g(x, y) = 6$:

The table below is constructed using nearby values for x and calculating corresponding values for y using $y = 6 - 2x$ (that is, $g = 6$ solved for y).

	x	y	$f(x, y)$	classification
checkpoint $(x_0 - 0.1)$	2.209	1.582	19.859	
critical point (x_0)	2.309	1.382	19.749	constrained minimum point
checkpoint $(x_0 + 0.1)$	2.409	1.182	19.859	

7. a. $450c + 400w = 80{,}000$ dollars gives the budget constraint where c is the number of units of capital and w is the number of units of worker hours

   b. The constrained system:
   $$\begin{cases} \text{optimize } P(c, w) = 100c^{0.6}w^{0.4} \\ \text{subject to } g(c, w) = 450c + 400w = 80{,}000 \end{cases}$$

   The four partials derivatives:
   $$P_c = 60c^{-0.4}w^{0.4}$$
   $$P_w = 40c^{0.6}w^{-0.6}$$
   $$g_c = 450$$
   $$g_w = 400$$

   The Lagrange system is:
   $$\begin{cases} 60c^{-0.4}w^{0.4} = 450\lambda \\ 40c^{0.6}w^{-0.6} = 400\lambda \\ 450c + 400w = 80{,}000 \end{cases}$$

   Solving the Lagrange system yields the point $(106.667, 80, 9507.213)$.

   (For algebra help see the bulleted section at the end of the main solution.)
   capital: 106.667 units
   worker-hours: 80 units

   • Algebra help for solving the Lagrange system:
   $$\begin{cases} 60c^{-0.4}w^{0.4} = 450\lambda & [1] \\ 40c^{0.6}w^{-0.6} = 400\lambda & [2] \\ 450c + 400w = 80{,}000 & [3] \end{cases}$$

   Rewrite equations [1] and [2]:
   $$\begin{cases} \dfrac{60c^{-0.4}w^{0.4}}{450} = \lambda & [1a] \\ \dfrac{40c^{0.6}w^{-0.6}}{400} = \lambda & [2a] \end{cases}$$

Setting LHS of equation [1a] equal to LHS of equation [2a]

$$\frac{60c^{-0.4}w^{0.4}}{450} = \frac{40c^{0.6}w^{-0.6}}{400}$$

$$400(60c^{-0.4}w^{0.4}) = 450(40c^{0.6}w^{-0.6})$$

$$c^{-0.4}w^{0.4} = 0.75c^{0.6}w^{-0.6}$$

$$w^{0.4}w^{0.6} = 0.75c^{0.6}c^{0.4}$$

$$w = 0.75c \qquad [4]$$

The Lagrange system can be reduced to:

$$\begin{cases} w = 0.75c & [4] \\ 450c + 400w = 80,000 & [3] \end{cases}$$

Substituting equation [4] into equation [3] gives:

$$450c + 400(0.75c) = 80,000 \qquad [5]$$

Solving equation [5] yields $c \approx 106.667$.
Substituting $c \approx 106.667$ into equation [4] gives $w = 80$.
Substituting $c \approx 106.667$ and $w = 80$ into equation [1a] gives $\lambda = 0.119$.
$P(106.667, 80) \approx 9507.213$

• Matrices are not used to solve this system because not all of the equations are linear. Verifying the point is a constrained maximum point by comparing output values of P at close points on the constraint $g(c, w) = 80,000$:

The table below is constructed using nearby values for c and calculating corresponding values for w using $w = 0.75c$ (that is, $g = 80,000$ solved for w).

	c	w	$P(c, w)$	classification
checkpoint $(c_0 - 0.4)$	106.267	80.45	9507.113	
critical point (c_0)	**106.667**	**80**	**9507.213**	constrained maximum point
checkpoint $(c_0 + 0.4)$	107.067	79.55	9507.113	

c. $\lambda = 0.119$ pipes per dollar was calculated as part of the solution to the Lagrange system in part b.
Increasing the budget by $1 would increase the maximum production run by approximately 0.1 pipe.

9. a. $f(a, b) = SSE = [273 - (55a + b)]^2 + [355 - (65a + b)]^2 + [447 - (75a + b)]^2 + [550 - (85a + b)]^2$

$f(a, b) = (273 - 55a - b)^2 + (355 - 65a - b)^2 + (447 - 75a - b)^2 + (550 - 85a - b)^2$

Alternate form: $f(a, b) = 702,863 - 236,730a - 3,250b + 20,100a^2 + 560ab + 4b^2$

b. first partials (by the Chain Rule):

$$\frac{\partial f}{\partial a} = 2(273 - 55a - b) \cdot (-55) + 2(355 - 65a - b) \cdot (-65)$$
$$+ 2(447 - 75a - b) \cdot (-75) + 2 \cdot (550 - 85a - b) \cdot (-85)$$
$$= -236,730 + 40,200a + 560b$$

$$\frac{\partial f}{\partial b} = 2(273 - 55a - b) \cdot (-1) + 2(355 - 65a - b) \cdot (-1)$$
$$+ 2(447 - 75a - b) \cdot (-1) + 2 \cdot (550 - 85a - b) \cdot (-1)$$
$$= -3250 + 560a - 8b$$

Solving the system of partials $\begin{cases} -236{,}730 + 40{,}200a + 560b = 0 & [1] \\ -3250 + 560a - 8b = 0 & [2] \end{cases}$.

Solving equation [2] for b yields $b = \dfrac{-3250 + 560a}{8}$

Substituting $b = \dfrac{-3250 + 560a}{8}$ into equation [1] gives

$$-236{,}730 + 40{,}200a + 560 \cdot \frac{-3250 + 560a}{8} = 0 \quad [3]$$

Solving equation [3] for a yields $a = 9.23$

Substituting $a = 9.23$ into $b = \dfrac{-3250 + 560a}{8}$ gives $b = 239.85$

Solution: $\begin{cases} a = 9.23 \\ b = 239.85 \end{cases}$

Minimum SSE: $f(9.23, 239.85) = \boxed{110.3}$

Determinant of second partial derivatives (each second partial is constant):

$$\begin{vmatrix} 40{,}200 & 560 \\ 560 & 8 \end{vmatrix} = 635{,}200$$

(The positive determinant and positive f_{aa} guarantee that the point found is a minimum.)

c. $d(x) \approx 9.23x + 239.85$ feet gives the distance necessary to stop a 3000 lb. car on dry pavement when the speed of the car is x mph, data from $55 \le x \le 85$.

11. a. The closed contours near 110°W in North America appear to occur in the rocky mountains so are most likely $\boxed{\text{relative maxima}}$.
  b. Taklimakan Desert
  c. The closed contour between 110°E and 120°E in Asia appears to occur in a mountainous area to the south-east of the Gobi Desert so most likely contains a $\boxed{\text{relative maximum}}$.